Achieving Digital Transformation Using Hybrid Cloud

Design standardized next-generation applications for any infrastructure

Vikas Grover

Ishu Verma

Praveen Rajagopalan

BIRMINGHAM—MUMBAI

Achieving Digital Transformation Using Hybrid Cloud

Group Product Manager: Preet Ahuja

Publishing Product Manager: Suwarna Patil

Senior Editor: Romy Dias

Technical Editor: Irfa Ansari

Copy Editor: Safis Editing

Project Coordinator: Ashwin Kharwa

Proofreader: Safis Editing

Indexer: Tejal Daruwale Soni

Production Designer: Prashant Ghare

Marketing Coordinator: Rohan Dobhal

First published: June 2023

Production reference: 1190623

Published by Packt Publishing Ltd.

Livery Place

35 Livery Street

Birmingham

B3 2PB, UK.

ISBN 978-1-83763-369-2

www.packtpub.com

I would like to dedicate this book to everyone who gave me an opportunity, and supported me, in working on emerging technologies and 5G architecture. I have been blessed to find like-minded people and while working with them, I never feel alone. Together, as a tech community, we have a huge impact on businesses and technologies.

– Vikas Grover

This book is dedicated to my father, Chhedi Lal Verma, who was an avid reader and inspired us all with his selfless dedication to everything he put his heart into.

– Ishu Verma

I dedicate this book to all the authors in the world. Writing to convey a message, sharpen the flow, and avoid repetition is immensely challenging, no matter what genre!

– Praveen Rajagopalan

Contributors

About the authors

Vikas Grover is a leading expert in open source technologies and has worked with a diverse range of clients from the finance and telecommunications sectors, providing innovative solutions in areas such as derivatives, credit cards, payments, telecom OSS/BSS, and the private cloud. He is a respected speaker at industry events. Vikas is dedicated to helping others succeed, guided by his personal motto, *"Helping others get what they want is the key to getting everything you want in life."*

Acknowledging the journey that led to the creation of his book, I express gratitude to my wife and two sons – Iraj and Aveer - for their unwavering support, as well as my co-authors, whose collaboration and teamwork were essential to bringing this book to life.

Ishu Verma is a technology advocate at Red Hat focused on emerging technologies such as edge computing, IoT, and AI/ML. He enjoys working with fellow open source hackers to work on ideas for next-gen open source technologies to benefit various industry verticals such as telco and industrial. Before Red Hat, Ishu worked at Intel and Wind River Systems on IoT gateways, embedded processors, platforms, and software.

He is a frequent blogger and speaker at open source and industry forums. Ishu resides in the Valley of the Sun, Arizona with his wife and two boys.

I would never have thought about co-authoring a book until my dear colleague Fatih Nar introduced me to Vikas Grover, so I'm grateful to them both for thinking of me for this book. Praveen Rajagopalan brings exceptional technical and business depth to this book.

I'm indebted to the open source community and particularly my Red Hat colleagues Rimma Iontel, Hanen Garcia, and William Henry for their work on the design patterns used in this book.

I'm also thankful to the Packt team for their guidance and support in bringing this book to fruition.

Finally, huge credit to my wife, Sandhya, and my boys, Sahil and Akul, for their support and understanding during the long hours spent researching, writing, and rewriting.

Praveen Rajagopalan has over 20 years of experience in the field of information technology and started his professional career as a DevOps engineer. Praveen is currently a customer engineer at Google Cloud, helping enterprises with their digital transformation journey, which includes cloud transformation and application modernization across many different industry verticals. Recently, he has found passion in helping Google Cloud's customers explore SaaS as a business model and helping them transform their applications to run as SaaS on Google Cloud. Praveen's passion is to solve core business-impacting problems and help enterprises become more agile and accelerate their growth. Praveen currently lives in Silicon Valley with his wife and daughter.

Not in my wildest dreams would I have imagined co-authoring a book. I reflect back on my professional career with profound gratitude and thank the Almighty and his grace for the current state of my life. Let me start by thanking my wife, Roshni, and daughter, Manasa, for being very supportive of this endeavor and encouraging me throughout the journey. Your cooperation and understanding, sacrificing winter holidays and weekends, helped me to focus and deliver on my commitments. I would also like to thank my co-authors, Vikas and Ishu, for their collaboration and knowledge sharing. I enjoyed working on my chapters and my research helped me to sharpen my knowledge. It also gave me an opportunity to reflect and reminisce on my professional experience.

About the reviewers

As a specialist in transformation projects, **Mario Mendoza** possesses extensive experience in a variety of business and technology areas.

As a team lead in Red Hat's Iberia solution architect team, he currently manages close relationships with first-line customers in order to help them adopt DevOps, hybrid, and multi-cloud architectures, AppDev and cloud-native solutions, as well as application modernization.

I'd like to thank all my colleagues and managers I've been learning from for more than 30 years, and all those companies who gave me an opportunity and had confidence in me to share their needs, accept challenges, and build successful solutions, with perseverance along the way.

And thank you, Yolanda, my wife, for your infinite patience during the long journeys, projects, and days when I left you, and our daughters, alone.

Sunny Goel is a senior delivery principal at Slalom who has 15 years of experience in architecting, designing, implementing, and deploying enterprise-level applications in hybrid and multi-cloud environments for customers across industries. He is an ex-AWSer and has helped customers globally to migrate and modernize their workloads on the AWS platform. He holds a B.Tech in computer engineering from Kurukshetra University. He is a multi-cloud certified specialist who loves to explore new technologies. He is an active contributor to open source projects focused on cloud-native and observability services. He is also a trusted advisor who collaborates effectively with diverse stakeholders, from C-level executives to developers across multiple teams.

I'd like to thank my family – especially my wife, who pushed me to pursue this opportunity and supported me throughout while I was juggling multiple things on the professional front. It was a great learning experience for me. Also, I'm deeply grateful to all the people and communities out there who are working tirelessly to produce amazing content for the tech community to learn about cloud computing, DevOps, and security-related topics.

Valentina Rodriguez Sosa is a Principal Architect at Red Hat, focused on OpenShift and container adoption. She helps customers to achieve their modernization and adoption goals by creating solutions and patterns that can be replicated across any organization. She has over 16 years of experience across various companies and organizations, from small start-ups to 600,000-employee technology companies, defining system architectures and developing enterprise software. She also has a Master's in Computer Science and is pursuing an MBA with certifications in Kubernetes, the cloud, cloud-native, the Spring Framework, and best practices for software development.

Table of Contents

3

Provisioning Infrastructure with IaC   47

4

Communicating across Kubernetes   75

Part 2: Design Patterns, DevOps, and GitOps

5

6

7

Preface

An organization's desire to optimize its IT infrastructure is the key driver behind the adoption of hybrid cloud and the rapid growth of the cloud computing industry. Hybrid cloud combines the benefits of both public and private clouds, giving organizations the flexibility to choose the right environment for their specific needs.

The main reasons for hybrid cloud adoption include flexibility, scalability, and security. With hybrid cloud, organizations can easily move workloads between public and private clouds, allowing them to scale resources up or down as needed. Additionally, hybrid cloud provides improved security by allowing organizations to keep sensitive data in a private cloud while still taking advantage of the cost savings and scalability of the public cloud.

Analysts have predicted that the hybrid cloud market will continue to grow, with Kubernetes playing a significant role in its success. Reports have stated that Kubernetes will become the de facto standard for managing cloud-native applications and that the hybrid cloud market will benefit greatly from its adoption. The ability to seamlessly manage and orchestrate workloads across multiple cloud environments using Kubernetes is a key driver of the growing adoption of hybrid cloud.

Hybrid Cloud Architectural Patterns provides a comprehensive guide for designing and implementing hybrid cloud solutions. The authors explore the key concepts, best practices, and real-world examples of using hybrid cloud, focusing on using Kubernetes to manage and orchestrate workloads.

Who this book is for

This book is written for architects, engineers, and developers in the telecom across different industry verticals and particularly the telecom industry who are navigating the complexities of using hybrid cloud to deploy solutions like 5G network. This comprehensive guide offers insights and practical design guidelines for adapting and deploying cloud native technologies such as service mesh and Kubernetes.

This book covers a wide range of topics related to hybrid cloud solutions, including the latest developments in the telecom industry like 5G rollouts. It provides real-world examples and best practices using use cases from telecom as well as other industries for designing and deploying successful hybrid cloud infrastructure. In addition, the book offers valuable insights into the challenges of building a hybrid cloud solution in a rapidly evolving technological landscape.

What this book covers

Chapter 1, Adopting the Right Strategy for Building a Hybrid Cloud, provides an introduction to hybrid cloud, common misconceptions of hybrid cloud, and the 6-R Framework and its benefits, to address different kinds of workloads and tools for adopting hybrid cloud.

Chapter 2, Dealing with VMs, Containers, and Kubernetes, explains the different types of hypervisors, the anatomy of containers, different distributions of Kubernetes, and how they help with hybrid clouds.

Chapter 3, Provisioning Infrastructure with IaC, explains the methodologies to provision and operate the hybrid cloud infrastructure.

Chapter 4, Communicating across Kubernetes, focuses on communication design patterns and discusses technologies that operate at Layer 3 and Layer 7 that make distributed and decoupled applications working in different clusters work together.

Chapter 5, Design Patterns for Telcos and Industrial Sectors, discusses 5G Core, the 5G **radio access network (RAN)**. We also talk about industrial edge patterns, **operations support system (OSS)/business support system (BSS)** patterns, and management solutions.

Chapter 6, Securing the Hybrid Cloud, explains core security principles and components of security in the hybrid cloud, network security, data protection, and compliance and governance.

Chapter 7, Hybrid Cloud Best Practices, explains overall best practices around implementing the appropriate architecture for the appropriate use case and goes into detail about developing the architecture based on hybrid cloud guidelines. It covers a range of topics from the economics to the pitfalls of the hybrid cloud.

To get the most out of this book

Software/hardware covered in the book	Operating system requirements
Kubernetes CLI (kubectl)	Linux
Kubernetes	Linux
Ansible	Linux
ArgoCD	Linux

You can install kubectl by going to `https://kubernetes.io/docs/tasks/tools/`.

You will also need Docker, and that can be installed by going to `https://docs.docker.com/get-docker/`.

Minikube is another tool that makes it easy for you to try and learn Kubernetes. You can install it by going here: `https://minikube.sigs.k8s.io/docs/start/`.

Download the color images

We also provide a PDF file that has color images of the screenshots and diagrams used in this book. You can download it here: `https://packt.link/pgupD`.

Conventions used

There are a number of text conventions used throughout this book.

`Code in text`: Indicates code words in text, database table names, folder names, filenames, file extensions, pathnames, dummy URLs, user input, and Twitter handles. Here is an example: "Mount the downloaded `WebStorm-10*.dmg` disk image file as another disk in your system."

A block of code is set as follows:

```
required_providers {
  aws = {
    source  = "hashicorp/aws"
    version = "~> 4.16"
  }
}
```

When we wish to draw your attention to a particular part of a code block, the relevant lines or items are set in bold:

```
- name: Install all security and critical updates without a scheduled
task
  ansible.windows.win_updates:
    category_names:
      - SecurityUpdates
      - CriticalUpdates
```

Any command-line input or output is written as follows:

```
dnf update -y
dnf groupinstall "Development Tools" "Development Libraries" -y
```

Bold: Indicates a new term, an important word, or words that you see onscreen. For instance, words in menus or dialog boxes appear in **bold**. Here is an example: "Select **System info** from the **Administration** panel."

> **Tips or important notes**
> Appear like this.

Get in touch

Feedback from our readers is always welcome.

General feedback: If you have questions about any aspect of this book, email us at `customercare@packtpub.com` and mention the book title in the subject of your message.

Errata: Although we have taken every care to ensure the accuracy of our content, mistakes do happen. If you have found a mistake in this book, we would be grateful if you would report this to us. Please visit `www.packtpub.com/support/errata` and fill in the form.

Piracy: If you come across any illegal copies of our works in any form on the internet, we would be grateful if you would provide us with the location address or website name. Please contact us at `copyright@packt.com` with a link to the material.

If you are interested in becoming an author: If there is a topic that you have expertise in and you are interested in either writing or contributing to a book, please visit `authors.packtpub.com`.

Share your thoughts

Once you've read *Achieving Digital Transformation Using Hybrid Cloud*, we'd love to hear your thoughts! Scan the QR code below to go straight to the Amazon review page for this book and share your feedback.

`https://packt.link/r/183763369X`

Your review is important to us and the tech community and will help us make sure we're delivering excellent quality content.

Download a free PDF copy of this book

Thanks for purchasing this book!

Do you like to read on the go but are unable to carry your print books everywhere? Is your eBook purchase not compatible with the device of your choice?

Don't worry, now with every Packt book you get a DRM-free PDF version of that book at no cost.

Read anywhere, any place, on any device. Search, copy, and paste code from your favorite technical books directly into your application.

The perks don't stop there, you can get exclusive access to discounts, newsletters, and great free content in your inbox daily

Follow these simple steps to get the benefits:

1. Scan the QR code or visit the link below

https://packt.link/free-ebook/9781837633692

2. Submit your proof of purchase

3. That's it! We'll send your free PDF and other benefits to your email directly

Part 1:
Containers, Kubernetes, and DevOps for Hybrid Cloud

The book is focused on hybrid cloud, and the first part of the book is specifically dedicated to exploring topics that will help you to build a foundation for a hybrid cloud strategy that is effective and efficient. It delves into the use of containers and Kubernetes to manage application deployment. These four chapters provide a comprehensive understanding of hybrid cloud and its components, making it a valuable resource for anyone interested in the field.

Here is a list of the chapters that will be included in this section:

- *Chapter 1, Adopting the Right Strategy for Building a Hybrid Cloud*
- *Chapter 2, Dealing with VMs, Containers, and Kubernetes*
- *Chapter 3, Provisioning Infrastructure with IaC*
- *Chapter 4, Communicating across Kubernetes*

1
Adopting the Right Strategy for Building a Hybrid Cloud

Cloud adoption brings benefits in the areas of developer productivity, cost, business agility, and innovation. By now, most organizations have some cloud footprint. But every organization is not able to reap maximum rewards from cloud adoption.

As organizations progress on their cloud adoption journey, they realize that each cloud brings its own strengths and weaknesses and some of the applications need to be in their own private data center or in multiple clouds.

With various public cloud providers and computing and delivery models, the cloud seems to bring limitless options when defining architecture. As an IT leader, you can easily get overwhelmed with design options to drive significant rewards from the cloud.

Your business and technical requirements can surely guide you to make design decisions, but with ever-changing needs, unforeseen future demands, and security and control requirements, many organizations choose to go with a bit of both worlds – public and private cloud – and are adopting a hybrid cloud.

In this chapter, we will cover the following topics to provide you with an overview of a hybrid cloud, including its benefits and use cases and the key benefits to consider while defining the hybrid cloud strategy for your organization:

- Exploring cloud computing – types and service delivery models
- Defining the hybrid cloud
- Hybrid cloud strategy
- Addressing compliance considerations
- Automating security measures
- Finding the right balance between public and private clouds

- Evaluating available tools and technologies
- Understanding the benefits of hybrid cloud computing

Exploring cloud computing – types and service delivery models

Cloud computing is a versatile technology that offers different types of services and consumption models. I will list the main types of cloud computing models and service delivery models here:

- **Cloud computing types**:

  - **Public Cloud**: Cloud services provided by a third-party provider over the internet that can be accessed by anyone who pays for them

  - **Private Cloud**: Cloud services that are dedicated to a single organization and are not shared with any other organizations

  - **Hybrid Cloud**: A combination of both public and private cloud services that work together as a single system

  - **Multi Cloud**: Using multiple cloud providers to fulfill different cloud computing needs

- **Service delivery models**:

  - **Infrastructure as a Service (IaaS)**: Cloud computing infrastructure (such as servers, storage, and networking) that is provided as a service to customers

  - **Software as a Service (SaaS)**: Cloud-based applications that are provided as a service to customers and are accessed over the internet

  - **Platform as a Service (PaaS)**: A cutting-edge platform that empowers developers to create, evaluate, and launch applications without the need to manage complex infrastructure

Here is an illustration of the cloud computing model and the service delivery model:

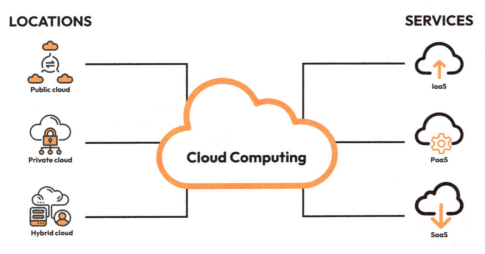

Figure 1.1 – Cloud computing model and service delivery model

The different cloud computing and cloud service delivery models offer different levels of performance, security, and cost-effectiveness. The public cloud model and the SaaS model are undoubtedly the most popular and widely adopted cloud computing and service delivery models, respectively. The following are the advantages of the public cloud and SaaS service model:

- Scalability
- Cost effectiveness
- Auto updates and reduced maintenance
- Flexibility

Organizations of all sizes and industries appreciate the convenience of adjusting their resources based on demand and only paying for what they use.

Leading public cloud service providers and SaaS offerings such as **Amazon Web Services** (**AWS**), Microsoft Azure, **Google Cloud Platform** (**GCP**), and Salesforce, respectively, have seen significant growth in recent years, catering to the needs of small start-ups and large enterprises alike.

However, it's important to consider that both models come with their fair share of drawbacks, and depending on an organization's background and goals, there can be differing views on the cloud.

While some visionary leaders are confident in the cloud's potential and are willing to invest heavily to offset rising cloud costs through product growth, others see cloud costs as a significant threat to their company's sustainability. For them, the fear of losing valuation due to soaring cloud expenses is a constant worry.

When approached with the right strategy, the cloud can offer numerous benefits to organizations. Not only does it enable better management of IT costs but it can also promote business growth by streamlining automation and reducing time to market.

However, it's important to note that each organization's approach to cloud adoption may vary in order to achieve the best results. One common mistake is when IT management treats cloud adoption as simply another IT system upgrade or uses a one-size-fits-all approach.

Designing a successful cloud infrastructure requires careful planning and foresight. While we can't always predict future needs, it's crucial to design with agility in mind, allowing applications to adapt quickly to meet evolving client demands while still maintaining cost-effectiveness.

Defining the hybrid cloud

The public cloud's pay-as-you-go offerings can be enticing, but for various reasons such as security, intellectual property, and cost of ownership, organizations need to preserve their existing workloads and assets in private data centers.

These factors, along with the growing use of edge computing, make a hybrid cloud a necessary solution to meet current and future needs. But before diving into the hybrid cloud, it's important to dispel a common misconception.

Some organizations may run certain workloads on public cloud providers such as AWS, GCP, or Azure while running other workloads in their private data centers. While these workloads are running in both public and private cloud environments, this hosting setup is not truly a hybrid cloud. Instead, these environments are isolated silos.

A true hybrid cloud is about creating a consistent platform across multiple environments.

According to the *Gartner Glossary*, "*hybrid cloud computing refers to policy-based and coordinated service provisioning, use, and management across a mixture of internal and external cloud services.*"

The **National Institute of Standards and Technology** (**NIST**) defines hybrid cloud as "*the cloud infrastructure [which] is a composition of two or more distinct cloud infrastructures (private, community, or public) that remain unique entities but are bound together by standardized or proprietary technology that enables data and application portability (e.g., cloud bursting for load balancing between clouds).*" [Source: NIST SP 800-145]

In our words, a hybrid cloud is a pool of computing power, storage, and services that is available from multiple environments, including the following:

- More than one public cloud
- More than one private cloud
- Private and public cloud combination

The ratio of consumption between private and public clouds varies based on the industry you're in, and it evolves as per compliance needs and time.

Variations in the hybrid cloud – homogeneous and heterogeneous

Variations in the hybrid cloud are entirely possible. You can have the following:

- Homogeneous hybrid cloud
- Heterogeneous hybrid cloud

Choosing between these two is based on your needs and strategy.

When you run the same technology stack in both public and private clouds, it's homogeneous. Traditionally, a single software vendor, such as Red Hat or VMware, provides a software stack including the operating system, hypervisor, and management layers for both clouds.

But when you run different components from different vendors and integrate them, that would be a heterogeneous cloud. You would have public cloud providers, such as AWS and Azure, and private cloud capabilities would come from Red Hat, VMware, and so on, and would be integrated with the public cloud at different levels.

Both come with pros and cons. While homogeneous can bring ease of usage but vendor lock-in, heterogeneous can provide more control and some complexity. You will want to consider various aspects before choosing which one you would like to implement:

- How much control you would like to have architecturally
- IT skills in your organization
- Cost and resources

Ultimately, it's about the appropriate platform for your respective applications. Organizations are looking at the cloud from economics, security, and use case points of view.

It is not always possible to move every workload to the public cloud. Organizations are also mindful of losing control of data and applications. Also, moving everything to the public cloud would mean that organizations are limited to the capabilities of the public cloud and costs can go out of control.

A hybrid cloud, on the other hand, will have resources distributed across on-premises, private, and public cloud environments.

This means a balanced approach where organizations get the speed and scale of the public cloud with the security and cost-effectiveness of the private cloud.

Because of the benefits the hybrid cloud brings and organizations' requirements, we are witnessing offerings by the public cloud that accommodate existing investments in private data centers. Some examples include VMware Cloud on AWS, VMware on Azure, and SAP on Google Cloud.

Many enterprises want to port on-premises virtual machines to the public cloud. The following diagram, taken from AWS, is a high-level component architecture reflecting VMware Cloud on AWS:

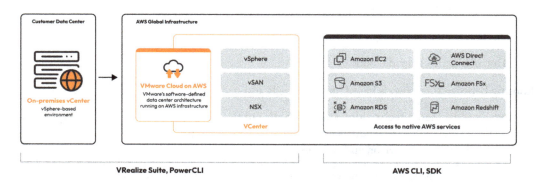

Figure 1.2 – VMware architecture on AWS

Not only that, but public cloud providers have also built extensions that push cloud solutions to organizations' private data centers. For example, AWS Outposts provides a hybrid experience by extending the AWS infrastructure, services, and APIs to on-premises in a fully managed offering. Google Anthos, Azure Stack, are also similar offerings by cloud providers:

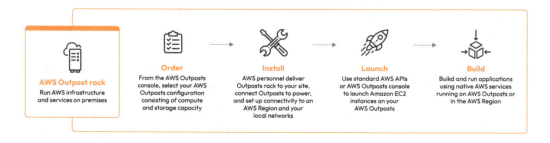

Figure 1.3 – AWS Outposts for on-premises

On a heterogeneous hybrid cloud, we have seen technologies and platforms from Red Hat, Pivotal Cloud Foundry (acquired by VMware), Nutanix, and so on that provide abstraction layers and create hybrid environments across distinct technology platforms.

Making public and private clouds work together should not be an afterthought. Create a comprehensive plan that accounts for applications, automation, management, and technology stack.

Increasing footprint

In terms of stats, Gartner reckons that *"by 2026 cloud spending is forecasted to exceed $1 trillion USD worldwide, exceeding all other IT markets. The drivers for this healthy state of affairs include cloud variations (such as hybrid IT and multiclouds. By 2020, 75% of organizations will have deployed a multicloud environments), which are now at the center of where the cloud hype currently is."*

Enterprises adopt different clouds because no one size fits all:

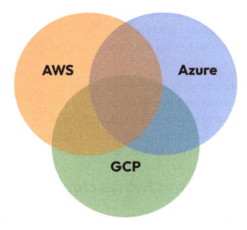

Figure 1.4 – Multi-cloud adoption by enterprises

This also brings in interesting questions that every architect and developer working in enterprise should try and find an answer to. Do you know which different clouds are adopted in your company and what percentage of applications in your organizations are portable and can run almost anywhere?

From various trends and reports, it is clear that companies looking at their future are choosing the hybrid cloud to take advantage of existing on-premises investments and the public cloud's scalability. A hybrid cloud provides the best of both worlds by giving the control and ability to innovate. This can be complex and thus organizations need a strategy to determine which workloads will reside where.

Hybrid cloud use cases

Hybrid cloud has emerged as a popular solution for organizations looking to balance the benefits of public and private clouds while addressing the data security requirements, compliance needs for regulated applications, and performance and computing needs for applications running at remote edge locations. Here are four use cases that showcase the versatility and flexibility of the hybrid cloud in different industries:

- **Security**: A government agency uses a hybrid cloud approach to store sensitive national security data on a private cloud for maximum security while utilizing the public cloud for cost-effective data storage and processing for non-sensitive data.

- **Proprietary Technology**: A technology company uses a hybrid cloud approach to store and manage its proprietary software on a private cloud for maximum security and control while utilizing the public cloud for cost-effective development and testing. For example, financial service companies manage trading platforms on the private cloud for maximum control while using the public cloud for running simulations and back-testing algorithms.

- **Competitive Edge**: A retail company uses a hybrid cloud solution to store critical sales and customer information on a private cloud for security and compliance while utilizing the public cloud for real-time data analysis to gain a competitive edge by offering personalized customer experiences and insights.

- **Telecom**: A telecommunications company uses a hybrid cloud approach to securely store sensitive customer information on a private cloud while utilizing the public cloud for real-time data processing and analysis to improve network performance and customer experience. This approach helps the company maintain a competitive edge in the telecom sector by providing a superior network experience to its customers.

Understanding the benefits of hybrid cloud computing

A hybrid cloud provides a flexible solution. Many organizations have embraced and adopted the hybrid cloud. If we take an example of a cable company, Comcast (the world's largest cable company), as per a technical paper published by Comcast for SCTE-ISBE, Comcast serves tens of millions of customers and hosts hundreds of tenants in eight regions and three public clouds. This is a great testimony of using a hybrid cloud for mission-critical workloads that need to run at scale.

Hybrid cloud is more popular than ever and some of the reasons that organizations are adopting a hybrid cloud are as follows:

- **Time to market**: With choices available to your IT teams to leverage appropriate resources as needed by use case, new applications and services can be launched quickly.

- **Manage costs**: Hybrid cloud helps you with optimizing and consuming resources efficiently. Make use of your current investments in existing infrastructure and when needed to scale, burst the workloads in the public cloud.

- **Reduced lock-in**: Going into the cloud may be appealing, but once in and when costs start to rise and eat the bottom line of the organization, it would be another costly proposition to reverse-migrate some of your applications from the public cloud. A hybrid cloud allows you to run anywhere and reduces your lock-in.

- **Gaining a competitive advantage**: In the competitive world of business, relying solely on public cloud technologies can put you at a disadvantage. To stay ahead of the competition, it's important to maintain control over and ownership of cutting-edge technologies. This way, you can build and grow your business in an increasingly competitive environment.

For example, consider a telecommunications company that offers mobile services. By investing in and owning the latest advancements in wireless technology, the company can differentiate itself from other providers and offer a more seamless, high-speed network experience to its customers. This could result in more loyal customers and a stronger market position, giving the company a competitive edge in the telecommunications industry.

- **Flexibility**: With common operating systems and a hybrid cloud, you can run applications on any infrastructure or cloud.

A hybrid cloud is a great option when your organization is looking to benefit from the best of different computing worlds, and by adopting an open architecture, open source technologies, and vendor-agnostic solutions, you can increase your preparedness for hybrid and unseen future needs.

Hybrid cloud strategies

To benefit from a hybrid cloud, it's important to have consistency and standardization while using distinct combinations. This can be achieved through the following:

- **Abstraction**: Different clouds become hybrid when your applications are abstracted from underlying infrastructure and connectivity is seamless to a great degree.

- **Portability**: A hybrid cloud should offer portability across environments.

- **Unified management**: Enforcing policies at scale across different clouds and environments is important to ensure standardization and compliance. A hybrid cloud needs unified management, orchestration, and security.

Your applications can reap significant benefits from such a setup where UI/UX runs on a public cloud and applications and databases run on a private cloud to comply with security and compliance needs or to manage costs.

When setting up the strategy for a hybrid cloud, key things to consider include the following:

- **Operating system**: A consistent operating system across clouds acts as a foundation. It provides the ability to host, manage, and monitor applications anywhere using a single set of tools.

- **Application categorization and rationalization**: Build an inventory of applications and categorize them according to the functionality they serve. Determine what to do with these applications. In the upcoming sections, we will explore the R framework to categorize applications.

- **Automation**: An assembly line that functions without much intervention is a must to take full advantage of the cloud. The automated creation of test environments, continuous integration, and continuous delivery is a must to increase operational efficiency.

- **Data-driven approach**: Data has traditionally lived in data centers. In the digital era, your customers demand insights and experiences in real time, and thus computing needs to be where your data is. It's the next stage of digital transformation, which takes data closer to the users who consume and create it. Determine where you need a computing pool and design your hybrid cloud around your data needs.

- **Management**: To enforce policies and reduce operational overhead, unified management is strategic for a hybrid cloud.

- **Technology partner**: A skills gap is the biggest hurdle, and it is very hard to attract talent and fill the skills gap. By partnering with experienced software vendors, organizations can benefit from their best practices and deliver hybrid clouds.

We discussed setting up the strategy for a hybrid cloud so that organizations can get the best of both public and private clouds. Organizations choose a hybrid cloud to deliver agility and meet business demands. However, for some industries, compliance and regulations are the primary reasons for a hybrid cloud instead of a unique cloud provider. Let's also look at some of the compliance requirements in our next section.

Addressing compliance considerations

Regulations and compliance are driven by government and external factors. To comply with laws, policies, and regulations, organizations have to work to adopt and implement compliance controls.

With HIPAA in healthcare, PCI-DSS, and GLBA in financials, FISMA for US Federal Agencies, and HACCP for the food and beverage industry, you may need to factor compliance needs into your design and architecture.

The terms of your **service-level agreement** (SLA) should also be consistent with compliance rules, such as the following:

- Backup and data recovery
- Security responsibility
- Data retention limitations
- System availability and reliability

Public cloud vendors are responsible for the physical security of the infrastructure, but many organizations need to do their own firewalls and patching and manage access privileges.

With hybrid cloud solutions, organizations can get the best of both worlds, where the public cloud is for non-regulated data while regulated information lives in the private cloud. The control that the hybrid cloud provides mitigates the risks with data residence regulations.

Take an example from the healthcare industry, in which you need to comply with the **HIPAA** and other standards. Your goal should be to proactively prevent, detect, and mitigate security threats.

You should consider the following implementations for streamlined compliance:

- **Centralized web console**: A console to administer, patch, provision, and manage your operating environment.

- **Monitor and prevent configuration drift**: On-demand and periodic checks to determine any drift from the baseline of the system. You need up-to-date protection against new threats and vulnerabilities.

- **Automated security**: Implement a system based on HIPAA policies and conduct vulnerability scans, and generate reports.

We looked at how compliance and legal requirements can bring constraints that you need to consider during the design and implementation phase. Mostly, your compliance requirements are non-negotiable, and thus having strategy and tooling that makes it easier for your application teams to implement for compliance and audit teams to review for compliance is important. We will now look at the importance of automating security in your organization.

Automating security measures

When adopting a hybrid cloud, your workloads can deploy in a range of environments – bare metal, virtual machine, or public clouds – and thus security becomes more complex.

The growth of heterogeneous environments will increase the risk and make manual compliance monitoring almost impossible.

The application teams, infrastructure teams, and security teams of different environments work within their own boundaries and zones leaving a blind side to the vulnerabilities.

With growing footprints and the nature of distributed systems and teams, automation is the only way to prevent inconsistent patching and configurations. Automation helps with the rapid implementation of continuous security and day 2 security operations.

Also, having an enterprise-wide security strategy helps. By bringing a consistent strategy, automation becomes easier and thus you can have an assembly line model where software is delivered at scale in a secure manner. By automatically patching the software, your software and software supply chain can be trusted.

Automation needs to come at different levels. Let's look at them:

- **Operating system (OS)**: Having a hardened OS as per compliance and performing patch management protects the OS from viruses, malware, and remote hacker intrusions. It is important to keep the OS safe by using techniques such as antivirus software, endpoint protection, patch updates, traffic monitoring, and firewalls, and by providing the least privileges.

- **Provisioning of systems**: System provisioning is a repeated task and is a great candidate for automation. **Integrated IT Service Management (ITSM)** – for example, ServiceNow – to provision systems in pre-defined secure ways by running playbooks is key to achieving automation.

- **Workflow management**: Workflows or pipelines can build a software factory where your applications have to pass security gates at the time of building. Before deployment and during packaging, your application components go through scanning and are key to DevSecOps.

You can start with iterative steps and start automating your daily tasks to secure your stack. Security at every step and every layer is important to keep your organization safe and mitigate your risk of misconfiguration and attacks. Now, let's look at how to enable your applications for adopting a hybrid cloud.

Finding the right balance between public and private clouds

The inventory and complexity of applications can make it hard to determine how and where to start your cloud migration process.

To take advantage of cloud capabilities and prepare your business to transform digitally, you need to have a good assessment in place for your workloads and come up with a decision matrix to decide the future of the workloads.

Having a framework can help you navigate through the complexities and come up with a blueprint for guidelines that your organization needs to follow.

Having a framework and migration factory, as depicted in the following figure, helps to realize a hybrid cloud in an accelerated way:

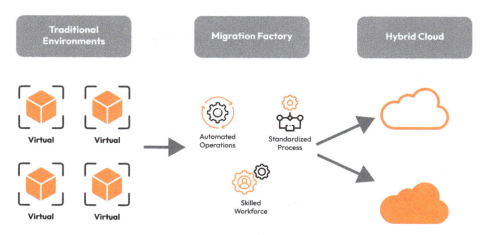

Figure 1.5 – Accelerate to a hybrid cloud by setting a migration factory

Using the **6-R framework** is a very effective way to determine the initial steps for cloud migration. Let's look at what each R means and stands for. The first two Rs are for Retire and Retain. These two strategies are for applications that may not be as strategic to the future of your organization. Let's look at these in a bit more detail:

- **Retire**: This is about retiring or decommissioning applications that are not needed, either now or in the near future. This can be looked upon as a great opportunity to identify and turn off certain applications that do not produce enough **Return on Investment** (**ROI**) for business. By retiring such applications, you can focus on services that are more needed and produce value.

- **Retain**: This is about maintaining the current footprint. It may be because you cannot get rid of it but also do not see any huge benefit by migrating such applications to the cloud. A certain portion of your portfolio will fall in this category because of security, ROI, or technical stack usage reasons.

Now that we have talked about two of the Rs that may address your non-strategic applications, let's look at the other four Rs and understand them in a bit more detail:

- **Rehost/Relocate**: The most commonly used strategy in organizations is rehosting. Even prior to the cloud, application owners and IT teams face certain roadblocks with current platforms because of cost or technical gaps and thus end up rehosting. This can be considered a simple migration that can bring significant benefits. It is also known as **lift and shift**. As the name implies, you lift/export your application from the current platform and deploy it on a new platform and make an immediate impact, and get ROIs.

  A few examples could be migrating your on-premises virtual machine to VMware on Cloud or to KubeVirt (KubeVirt makes it possible to run a virtual machine in a Kubernetes-managed container platform).

  Rehosting may not turn your applications cloud-native or provide benefits as replatforming/refactoring does, but given less resistance and friction, the cost is less and returns are realized quickly.

  Also, relocating (also known as **hypervisor-level lift and shift**) refers to the process of moving infrastructure to the cloud without the need to purchase new hardware, rewrite apps, or modify existing operations. This term is commonly used in the context of the VMware Cloud on AWS offering.

- **Replatform**: This can be looked upon as a further add-on to rehosting. For some applications, it is important to make additional optimizations and perform some tweaking and coding to get benefits from cloud capabilities such as elasticity, scale, self-healing, and so on.

- **Refactor**: This strategy is more fitting when certain applications are in need of extensive improvements to serve performance, availability, and reliability. Application teams have to do extensive design thinking and come up with an architecture that adheres to new non-functional requirements. This can be a time-consuming task and yet the most beneficial strategy, and it needs skill sets and expertise to take advantage of cloud-native capabilities.

- **Repurchase:** The last strategy is about moving on from existing vendors or technology and adopting new vendors. It means terminating your existing subscriptions and licenses for cost, security, or technical reasons – for example, giving up your on-premises **Customer Relationship Manager (CRM)** system to adopt a cloud-based SaaS from Salesforce or Workday. Another example is moving or reducing the usage of proprietary databases and adopting cloud-based databases.

The following table is a quick summary of the 6-R framework and how each strategy impacts time and costs and brings business benefits:

	6-R Framework	Time To Realize	Migration Costs	Operating Costs	Business Benefit
As Usual	**Retire** (Retire Workload)	•	•		
As Usual	**Retain** (Existing Hardware & Platform)			•••••	
Modernize	**Rehost** (New Hardware, Existing Platform)	••	••	•••••	••
Modernize	**Replatform** (As VMs)	••	•	•••	••••
Modernize	**Refactor** (As Containers)	••••	•••	•	•••••
Modernize	**Repurchase** (As Containers)	•••	•••	••	••••

Figure 1.6 – 6-R framework and benefits

We talked about the 6-R framework, which could be very handy to determine the fate of your applications and your approach toward them. It is not meant to be mutually exclusive and you can use or customize this framework as your circumstances demand. Let's look at different tools and technologies that could help in implementing the 6-R framework.

Evaluating available tools and technologies

Although clouds offer comparable functionalities to a certain degree, they have distinct characteristics. As each cloud, whether public or private, operates independently, your company's IT infrastructure may face compounded challenges due to the variety of instances, networks, and storage types across different clouds.

It is practically not possible for your team, which is trained and delivering solutions on one cloud, to efficiently translate their skills into another cloud. Thus, we see organizations hiring different team members from different backgrounds and experiences to manage clouds such as AWS, Azure, Google, and private clouds.

As an enterprise, your teams are trying to make the most out of your cloud subscription. It is also in the interest of your public cloud provider to have you use all of their offerings. However, the goal should be to get the best out of the different cloud subscriptions by making them work together.

The expectations from your tenants would be to be able to request cloud resources and manage user permissions and automated controls. The tenant can request different resources at different layers, as depicted in the diagram:

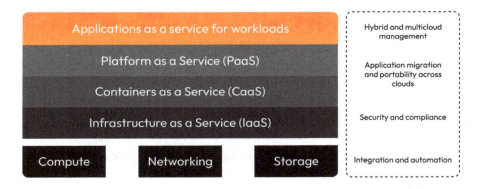

Figure 1.7 – Everything as a service

You need to look at certain characteristics to make a hybrid cloud a reality:

- **Common platform and operating environment**: A common operating environment is needed so that when users turn toward any cloud, they have a uniform experience at the platform and operating level. This will allow users to connect and manage applications in a streamlined manner.

- **Automation**: In a hybrid cloud environment, automation is crucial for achieving consistent and efficient management of both public and private cloud infrastructure. Cloud-agnostic tools such as Puppet, Chef, and Ansible provide IT teams with the ability to automate infrastructure configuration, application deployment, and ongoing management, regardless of the underlying cloud provider. These tools help organizations to standardize their operations, reduce manual errors, and ensure that their infrastructure and applications are secure, scalable, and highly available. Furthermore, when combined with GitOps, cloud-agnostic tools can help organizations to achieve a Git-centric approach to infrastructure as code, which enables them to manage their infrastructure and applications through a single source of truth and automated workflows. This provides a clear and consistent approach to managing their infrastructure, while also allowing them to take advantage of the benefits of both public and private clouds

- **Implement comprehensive security**: Security is complex and challenging. While the ultimate goal should be to secure at every layer, the approach should be to simplify security management. When your environments and infrastructure differ, applying the same security policy, applying patches, and changing management in different clouds becomes tedious. It would be ideal to have one tool that spans across multiple clouds. Acquiring tools to manage security and patches at a centralized and granular level across infrastructure will help accelerate cloud adoption. One such tool is OpenSCAP.

  OpenSCAP, a comprehensive open source initiative, offers a robust suite of tools for seamless implementation and enforcement of **Security Content Automation Protocol** (**SCAP**) standards, as diligently maintained by NIST.

  OpenSCAP performs vulnerability scans and validates security compliance content to generate reports. It is a great solution for fast and repeatable security.

- **Unified management**: A single control plane to manage the life cycle of multiple clusters agnostic to the underlying platform will be used by teams to create resources across clusters. Industry leaders in hybrid cloud management include Microsoft, Red Hat, and VMware. This provides the ability to deploy applications from different sources and have a consistent experience across all clusters, manage risk and apply policies for security, and maintain governance.

- **Policy and governance**: Policy and governance play a crucial role in the success of a hybrid cloud strategy. A well-defined set of policies and governance frameworks helps organizations to effectively manage security, compliance, and resource allocation across multiple cloud environments. The policies need to be flexible enough to adapt to changing business requirements while ensuring that the data and applications remain secure. The governance framework helps in defining roles, responsibilities, and decision-making processes, leading to better alignment and coordination between different teams. Additionally, a robust governance framework ensures that the hybrid cloud strategy is aligned with the overall business objectives and goals, leading to better cost optimization, risk mitigation, and overall performance. In conclusion, policy and governance form the backbone of a successful hybrid cloud strategy, and organizations must prioritize these aspects for seamless and efficient deployment and operation of hybrid cloud solutions.

- **Modernize applications**: Many such tools exist that help with migration to modernize applications. One such example is the open source tool, Konveyor. Konveyor (`https://www.konveyor.io/`) is a suite of tools that focuses on various use cases with the target platform of Kubernetes, and prime contributors to these tools are IBM Research and Red Hat with involvement from Microsoft. It is an open source **Cloud Native Computing Foundation** (**CNCF**) sandbox project. It includes a bundle of different tools that come under the umbrella of Konveyor. The following diagram from the Konveyor website does a pretty good job of depicting different Konveyor tools:

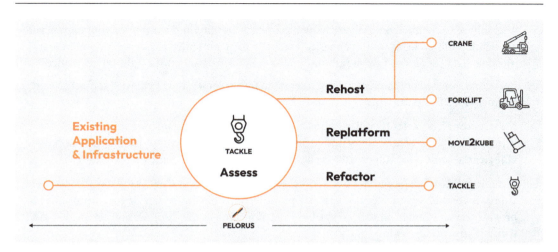

Figure 1.8 – Konveyor and tools

Let's briefly look at the various tools under the Konveyor umbrella:

- **Konveyor Move2Kube**: Replatforms applications to Kubernetes

- **Konveyor Crane**: Rehosts applications between Kubernetes clusters

- **Konveyor Tackle**: Assesses, prioritizes, and refactors applications

- **Konveyor Forklift**: Rehosts virtual machines to KubeVirt

- **Konveyor Pelorus**: Measures software delivery performance

You can go to the Konveyor website and look at demonstrations and source code and try these tools, which help to implement some of your 6R strategies.

In addition to the preceding, other solutions exist, such as the following:

- **Public cloud vendor offerings**: To maximize developer productivity, public cloud vendors came up with offerings such as AWS Outposts, Azure Stack, Google Anthos, and Google Cloud's operations suite (formerly Stackdriver), which allow you to build and deploy applications as normal both on-premises and on the public cloud.

- **Platform vendor offerings**: Various vendors offer solutions that span public and private clouds. Certain tools from vendors such as Scalr, Cisco Cloud Center, Red Hat OpenShift, and VMware Tanzu Application Service provide essential tooling in this area.

    As an example, Red Hat Advanced Cluster Management will bring the capabilities you need for your large hybrid environment. To control your clusters and applications from a single console, Red Hat Advanced Cluster Management plays a great role.

This solution provides comprehensive management, visibility, and control for your cluster and application life cycle, as well as enhanced security for your entire Kubernetes domain across multiple data centers and public clouds. It also offers compliance with industry regulations.

Because these are complementary and integrated technologies, they help with self-service and free up your IT departments.

- **Kubernetes**: Kubernetes (popularly known as **k8s** or **kube**) is a container orchestration platform. It is an open source technology and it came out of Google. Although initially developed by Google, the project for Kubernetes is currently under the stewardship of CNCF.

  It is the de facto standard and is declarative in nature, and also an ideal foundation for a hybrid cloud. It abstracts your workload from the underlying hardware. Thus, you can use k8s to provide the same environment everywhere and run containerized applications in any location without any modification.

  The flexibility to operate across any cloud and the elasticity of the cloud (as you can dynamically scale your Kubernetes clusters up or down based on workload demand) are why it is popular among organizations.

The tools and technologies that we looked upon come from various vendors, cloud providers, and the open source world. These tools help with management, provisioning, migration, optimizing, securing, and overall, helping you realize your hybrid cloud.

Summary

By now, you should have an overview of the hybrid cloud and its benefits. We also covered very useful and important tools that help with adopting the hybrid cloud in an accelerated manner.

We recommended that enterprises choose a common operating environment and modernize their applications to benefit from cloud capabilities. Similarly, using a common set of tools across on-premises and clouds can help you approach your public cloud as an extension of an on-premises data center.

By going to provided links/references and following instructions, you can download tools that can help you assess your application and define a migration strategy. You can also try open source tools such as Ansible for building your automation.

In the next chapters, we will learn about some vital technologies using use cases from 5G telecommunications.

Further reading

- Hybrid Cloud Glossary from Gartner: `https://www.gartner.com/en/information-technology/glossary/hybrid-cloud-computing`
- SAP on Google Cloud: `https://cloud.google.com/solutions/sap`

- VMware on AWS: `https://aws.amazon.com/vmware/`
- Azure VMware Solution: `https://azure.microsoft.com/en-us/products/azure-vmware/#product-overview`
- Red Hat OpenShift: `https://www.redhat.com/en/technologies/cloud-computing/openshift`
- Google Anthos: `https://cloud.google.com/anthos`
- VMWare Tanzu: `https://tanzu.vmware.com/tanzu`
- Azure Stack: `https://azure.microsoft.com/en-us/products/azure-stack`
- AWS Outposts: `https://aws.amazon.com/outposts/`

2
Dealing with VMs, Containers, and Kubernetes

In the last chapter, we went over the strategy for building a hybrid cloud. In this chapter, we will look at some of the key technologies that act as the foundation for hybrid clouds. Three of these technologies are virtualization, containers, and Kubernetes.

Before we start this chapter, I would like to differentiate between virtualization and the cloud. These two terms are often confused with each other because both relate to providing environments to host your applications. However, virtualization is a technology and the cloud is an environment. The cloud can involve bare metal machines, **virtual machines** (**VMs**), or containers and provides self-service access and dynamic resource pools. You use some or all of the technologies mentioned here and additional tools to build a cloud environment.

This chapter covers some key technologies and concepts that are the foundation of the cloud and the hybrid cloud.

We will cover the following topics to explain the technologies that help deliver a hybrid cloud:

- Introduction to VM and containers
- Anatomy of containers
- The differences between a VM and a container
- Container orchestration
- CI/CD on a hybrid cloud

Knowledge of VMs, containers, and Kubernetes will set the foundation for understanding the basics of the hybrid cloud. We will delve into the details of various topics, including containers, Docker, and building containers using Docker commands.

Introduction to VM and containers

It is very important to understand two significant forces in the cloud world – virtualization and containers. These technologies are similar in some ways and aim to provide efficiency, portability, and DevOps capabilities, but they do so differently and have some unique characteristics.

In later chapters, we will use examples and use cases from the telecoms and 5G world in the context of the hybrid cloud. The telecoms industry is evolving at a fast pace based on a **service-based architecture** (**SBA**). Virtualization, containers, and Kubernetes are key to the implementation of 5G and are at the heart of it. These technologies allow telecoms operators to bring new services to market without worrying about the underlying infrastructure/cloud.

Let's take some time to dig deep and understand these technologies.

VMs

Most of us run our workloads on VMs. By utilizing virtualization technology, specifically the hypervisor component, VMs are able to offer virtual compute, virtual storage, and virtual network resources.

Virtualization came into existence in the 1960s, and it started with the logical division of mainframe computers. Using the technological process of virtualization, a machine is made called a VM. VMs use hypervisors to emulate hardware and thus allow multi-tenancy by hosting multiple OSs and running them side by side.

VMs behave similarly to physical servers, but they utilize a hypervisor to isolate the underlying hardware resources and assign them to individual VMs. This enables VMs to partition and mimic complete OSs, databases, and networks.

The hypervisor is a key component of a VM. It handles the transactions between physical hardware and VMs and makes VMs portable because they can be moved between different machines, unlike traditional applications, which are tied to physical machines.

Another concept to understand is that hypervisors are broadly categorized into two categories, Type 1 and Type 2:

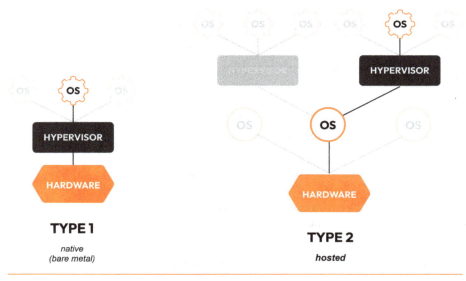

Tyep-1 and Type-2 hypervisors

Figure 2.1 – Type 1 and Type 2 hypervisors

The quickest way to understand the difference between Type 1 and Type 2 hypervisors is to look at the preceding diagram, but we explain the differences between Type 1 and Type 2 hypervisors in more depth here:

- **Type 1 Hypervisor**: This is also known as a **bare metal hypervisor**. It runs on a physical machine directly and does not load or use an OS. For this reason, it is the best-performing hypervisor as it has no OS or drivers to deal with.

  Furthermore, due to the absence of an intermediate OS layer, the vulnerabilities associated with the OS are eliminated. Type 1 hypervisors are widely recognized for their security. They rely on specialized hardware equipped with acceleration technology to efficiently handle demanding tasks such as memory management, storage, network resources, and CPU operations.

  Examples of Type 1 hypervisors are VMware ESXi and Microsoft Hyper-V.

- **Type 2 Hypervisor**: This is also known as a **hosted hypervisor** and it runs on an existing OS. For this reason, Type 2 hypervisors introduce a certain amount of latency.

  Also, because of the OS layer, operations such as managing memory, storage, network resources, and CPU calls are delegated to the OS and thus Type 2 hypervisors can run on a wide variety of hardware.

  Examples of Type 2 hypervisors are Oracle Virtual Box, Microsoft Virtual PC, VMware Workstation, and VMware Fusion.

Type 1 hypervisors are more often seen in production environments, while Type 2 hypervisors are more often seen in test environments and home labs. The following table lists Type 1 and Type 2 parameters:

Parameter	Type 1	Type 2
Performance	High	Medium
Execution	Runs on hardware	Runs on an OS
Security	High	Medium
Used in	Production environments	Mostly in non-production

Table 2.1 – Parameters of Type 1 and Type 2 hypervisors

In addition to Type 1 and Type 2 hypervisors, we also have a mix of both Type 1 and Type 2 hypervisors called **Kernel-based Virtual Machine (KVM)**.

KVM is a virtualization module integrated into the Linux kernel that enables the kernel to function as a hypervisor. KVM was introduced in Linux kernel version 2.6.20 and is considered a Type 1 hypervisor due to its direct integration with the OS. However, it is also categorized as a Type 2 hypervisor as it uses the Linux OS. This unique technology combines the benefits of Type 1 and Type 2 hypervisors, making it an intriguing option for virtualization in Linux-based environments.

KVM converts Linux into a Type 1 bare metal hypervisor and only one kernel (the Linux kernel) is used, as shown in the following diagram:

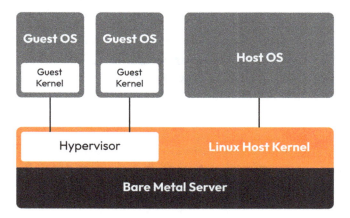

Figure 2.2 – KVM

KVM allows the kernel to act as a hypervisor.

Let's take a minute to look at a diagram of Type 1, Type 2, and KVM hypervisors next to each other:

Figure 2.3 – Different hypervisors

As you can see, KVM is part of Linux, and as long as you are running a Linux version released after 2007, you can implement KVM (by loading specific modules and drivers; please refer to the manual for your Linux distribution).

Containers

Containerization is similar to virtualization at the application level and provides a fully functional portable environment for applications. Container technology is different from virtualization in that containers share resources with the same underlying host OS, while a VM has its own OS packaged in it.

In the *The differences between VMs and containers* section, we will explain how virtualization differs from containerization. Fundamentally, a container is a process that has a limited and restricted view, making it isolated from the system. The containerized application is usually packaged into one file with all its dependencies (libraries, etc.) for easy migration between systems.

Containers share the same OS kernel and, thus, they are less resource-intensive than VMs. Abstracted from the underlying infrastructure, containers deliver portable and scalable applications. Containers work across distributed cloud architectures and are a top choice for cloud-native development. Containers play a vital role in removing the complexity involved in moving applications across environments and the cloud.

This is why the adoption of container technology is happening at a fast pace. Software vendors and traditional enterprises are adopting containers for the benefits they provide in the areas of DevOps, CI/CD, and portability.

We have looked at VM and containers. Let's now look at what makes containers possible by going into the anatomy of containers.

Anatomy of containers

Containers have been in existence for more than 10 years. Fundamentally, containers are part of Linux, or a feature of Linux, and they are processes that run on Linux. But these processes are isolated from other processes on the same host OS.

Containers are isolated because of some components, such as namespaces, cgroups, and SELinux. These components make containers secure and enterprise-grade. The following figure shows the components of a container:

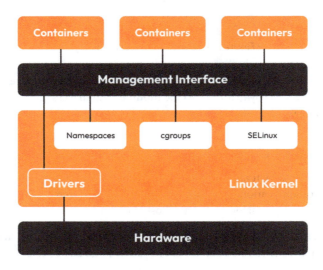

Figure 2.4 – Container components

Let's look at the components shown in the preceding diagram in more detail:

- **Namespaces**: Namespaces limit what the process can see and are created with syscalls. Namespaces are used by the Linux kernel to provide process isolation.

- **Cgroups**: Because containers run on a single host, it's always a worry that one or more containers can consume a large amount of resources, thus depriving other containers. cgroups ensure that each container is allocated its own resources – CPU time, network bandwidth, and system memory.

- **Seccomp**: This stands for **Secure Computing**, and its purpose is to restrict system calls that are allowed to be made by an application. Containers do not need to change various attributes of the host (for instance, modifying kernel modules), and thus syscalls such as `create_module` will be blocked.

- **SELinux**: SELinux applies policies and labels to define access controls for the applications, processes, and files, and hence secures container resources by keeping them separate.

Now that you understand the components that make containers secure and enterprise-grade, let's discuss Docker.

You may have come across Docker when discussing containers in your organization. The terms "Docker" and "container" are used interchangeably (even though containers existed before Docker was released).

I will also highlight that the company Docker is different from Docker the image format. Readers should understand that Docker, Inc. is the company and Docker is a product.

Docker has been a very popular project, with thousands of contributors in open source projects. The primary reason for the adoption of Docker is that it lets developers access container capabilities using commands, and it allows automation using APIs, improving productivity.

The Docker open source project was mainly contributed to by Red Hat, Docker Inc, Microsoft, IBM, Google, Cisco Systems, and Amadeus IT Group. The technology initially developed in the Docker project was later standardized, and its development was moved to the **Open Container Initiative** (**OCI**) open containers project, ensuring container interoperability between different platform providers.

About OCI and Docker

The OCI is an industry standard for container image formats and runtime specifications designed to ensure the interoperability and portability of containers across different platforms and tools. OCI was founded in 2015 by a group of industry leaders, including Docker Inc., to promote a common, open standard for containerization.

Docker, on the other hand, is a proprietary implementation of the OCI standard, providing a set of tools and services for developing, packaging, and deploying containerized applications. Docker has been a key player in the containerization ecosystem, helping to popularize the use of containers and contributing to the development of the OCI standard.

However, Docker also includes non-standard functionalities and extensions that are not part of the OCI standard. This can make it difficult to use Docker containers with other tools and platforms that adhere strictly to the OCI standard.

To address this issue, the OCI standard was developed as an open, vendor-neutral standard for containerization. The OCI standard provides a set of specifications for container images and runtime environments, ensuring that containers can be built, run, and managed in a consistent and interoperable way across different tools and platforms.

The OCI specifications define two key components:

- **The container image format (OCI image specification)**: This describes the layout and contents of a container image, including how the image is built, layered, and distributed.

- **The container runtime specification (OCI runtime specification)**: This defines how the container image is executed and how the container interacts with the host OS and other containers.

Overall, while Docker has played a significant role in the containerization ecosystem, the OCI standard represents a more open and standardized approach to containerization, providing greater interoperability and portability for containerized applications.

Moving forward, let's explore some of the fundamental terminology utilized in Docker:

- **Docker image**: A Docker image is source code and also has libraries and dependencies that code needs to run as an application. A Docker image, at runtime, is called a container.

- **Dockerfile**: Images are defined in text files with all the steps required to use them. These files are called Dockerfiles.

- **Docker container**: When a Docker image is actively deployed and operational, it is referred to as a container. Containers are dynamic, transient, and executable entities that host live content.

Let's put this terminology into use by creating a Docker image and running it:

1. First of all, install Docker on your machine by following the instructions on the Docker website (`https://docs.docker.com/get-docker/`). Follow the instructions on the site that correspond to the OS you have on your desktop/laptop, or use Buildah as a native alternative for Linux systems. You can refer to the instructions here. We will also see some Buildah commands in a later part of this section (`https://github.com/containers/buildah/blob/main/install.md`).

2. Now that Docker is installed and up and running, you can create a Dockerfile and run your Docker image.

   Create a folder with two files:

   - `Dockerfile`

   - `main.py`

   Update `main.py` and paste the following code:

   ```
   #!/usr/bin/env python3
   print("Docker is running")
   ```

   Edit `Dockerfile` with the following commands:

   ```
   FROM python:latest
   COPY main.py /
   CMD [ "python", "./main.py" ]
   ```

3. Now you can build the Docker image:

```
$ docker build -t python-test .
```

4. You are ready to run your code:

```
$ docker run python-test
```

As you can see, your application was packaged inside `Dockerfile` and is run as a container on your machine. You can have a reproducible environment, and you can run this container wherever you have Docker Engine. This is a very powerful capability.

As mentioned earlier, let's also look at Buildah.

Buildah is a command-line tool for building container images without requiring a Docker daemon or Dockerfile. Here are some basic commands for using Buildah:

- Create a new container:

```
buildah from <base-image>
```

This command creates a new container from the specified base image.

- Run commands in the container:

```
buildah run <container> <command>
```

This command runs the specified command in the container.

- Install packages in the container:

```
buildah run <container> yum install -y <package>
```

This command installs the specified package in the container.

- Copy files into the container:

```
buildah copy <container> <src-path> <dest-path>
```

This command copies files from the specified source path into the container at the specified destination path.

- Commit changes to the container:

```
buildah commit <container> <image-name>
```

This command commits the changes made to the container and creates a new container image with the specified name.

- Push the image to a registry:

```
buildah push <image-name> <registry>/<repository>: <tag>
```

This command pushes the specified container image to the specified registry with the specified repository name and tag.

Overall, Buildah provides a flexible and powerful tool for building container images without relying on Docker, and can be used in a variety of containerization workflows.

In this section, we covered the commands for both Docker and Buildah. Containers play a vital role in the cloud-native environment and have revolutionized software development, deployment, and release processes.

I trust that you found our discussion on containers and Docker informative and insightful. Moving forward, let's delve deeper into the distinctions between VMs and containers.

The differences between VMs and containers

We have looked into both VMs and containers. Both of them make multi-tenancy possible and increase resource utilization (CPU and memory). Now, we will look at what makes VM and containers different. Gaining a deeper understanding of VMs and containers can enable you to make informed decisions about which technology to use based on your organization's architectural and business objectives. By having a more thorough understanding of these technologies, you can make more informed choices about when and how to utilize them. The following diagram shows VMs and containers next to each other and you can see how they differ at a high level.

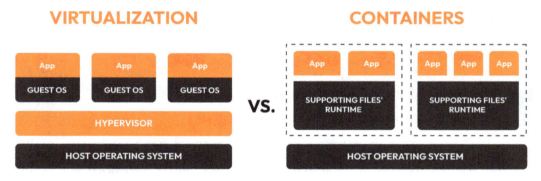

Figure 2.5 – The differences between VMs and containers

VMs are traditionally used for monolith workloads. VMs are less portable than containers because of the dependence they have on the OS, applications, and libraries. VMs are mostly used for the following:

- Running guest OSs based on the application needs on top of the host OS
- Hosting traditional legacy and monolith workloads
- Provisioning resources such as networks and servers (infrastructure resources)

On the other hand, containers are mostly used for microservices and are lightweight by nature, although legacy/monolith workloads can be run as well. The small footprint of containers allow them to be moved easily across public, private, hybrid, and multicloud environments. Containers are used for the following:

- Developing cloud-native applications
- Packaging microservices
- Running your workloads at scale across different IT environments

Containers also help with cost effectiveness and agility, and are a vital part of hybrid cloud strategies.

However, this was not an attempt to conduct a boxing match between these two technologies. The idea of explaining the differences is to help you choose the right technology for your requirements.

While virtualization may sound like it is at a disadvantage, keep in mind that virtualization technology has matured over many years and comes with proven resilience capabilities such as **Software-Defined Networking (SDN)**, live migration, high availability, and storage integration (the integration of storage with containers is still maturing).

We hope this gives you good comparison of VM and containers. In next section, we will look at what do we need to run containers at scale.

Container orchestration

When your organization is scaling up and has hundreds of containers, you need tooling to automate the deployment, management, and scaling of your containers.

Container orchestration is about tasks such as the following:

- Deployment and provisioning
- Allocation of resources
- Traffic routing
- Availability and fault tolerance
- Monitoring the health of containers
- Container communication/networking
- Scaling

Why do we need container orchestration?

Container orchestration is vital for the successful use of containers, especially when running containers at scale. Many container orchestration tools are available for container life cycle management. We will look into them in this section.

An orchestration tool will help you with the management of containers and with achieving agile delivery with DevOps. It will also help you manage your containers across different clouds.

Keep in mind that containers do not live for long; thus its important to consider the following:

- High availability
- Scalability
- Security
- Networking
- IP allocation
- Discovery

Many container orchestration tools exist, and you will find that among them, the de facto standard and most popular is Kubernetes (or you will notice that many distributions are a flavor of Kubernetes).

Kubernetes – a container orchestration tool

Kubernetes was developed and designed by Google engineers, and it was donated to the **Cloud Native Computing Foundation (CNCF)** in 2015. It is a popular open source platform and is highly extensible and portable.

Kubernetes provides a declarative, resource-centric REST API that allows developers and administrators to describe the desired behavior of a system. This means that instead of giving explicit instructions on how to achieve a certain state, users can simply define the desired state, and Kubernetes dynamically executes the necessary actions to achieve that state. This approach helps with the following:

- Scheduling
- Self-healing
- Horizontal and vertical scaling
- Service discovery
- Automated rollouts and rollbacks

The following diagram shows the master and worker nodes of Kubernetes:

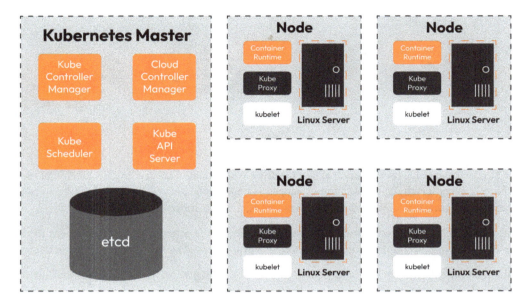

Figure 2.6 – Components of Kubernetes

The main components of Kubernetes are as follows:

- **Pod**: A group of one or more containers deployed is known as a Pod. The Pod is the smallest deployable unit with shared storage and network resources. Kubernetes manages Pods rather than managing containers. The most common scenario is that you run one container in a Pod. However, you can have more than one container in a Pod. As an example, in the case of Istio, a sidecar container that acts as a proxy runs next to your application container, both in the same Pod.

  Kubernetes allows the transparent failover from one Pod to another with no impact on the user.

- **Control plane**: In a Kubernetes cluster, the worker nodes are responsible for hosting the application containers, while the control plane manages and oversees the worker nodes and Pods. The control plane acts as the "brain" of the cluster, making global decisions and monitoring the overall health and state of the cluster. It maintains a database called `etcd`, which serves as a centralized store for storing the configuration and state information of the entire cluster.

- **kubelet**: A kubelet is an agent that runs on each node. When a Pod is started, a connection kubelet is required. The kubelet uses the container runtime to start the Pod, monitor its life cycle, and check for readiness.

To run containers in Kubernetes, an interface called a **container runtime interface** (**CRI**), which is based on gRPC, is used. The container runtime that implements the CRI can be used on a node controlled by the kubelet.

Figure 2.7 – CRI

With Kubernetes, it's all very declarative: you describe your configuration using YAML/JSON files. These files tell you everything: the location of images, storage, replicas (how many copies need to run), application configuration, CPU and memory resources, and lots more.

For the deployment of any container, Kubernetes automatically schedules the deployment to the cluster and takes into account various factors and constraints based on the provided specifications.

Container orchestration is needed in any container environment, including on-premises, and private and public cloud environments.

Let's look at other commercial container orchestration tools as well.

OpenShift

OpenShift is Red Hat's (now acquired by IBM) flagship product, and it is built on top of Kubernetes. In addition to what Kubernetes provides, OpenShift adds various enterprise-ready features that extend and secure Kubernetes.

OpenShift is 100% open source. It is designed for an open hybrid cloud strategy and provides consistency to your applications deployed in your data center, in your cloud, or on the edge.

The official Red Hat OpenShift site provides a lot of information and explains how OpenShift runs anywhere, and it also shows additional tooling from Red Hat that provides features and benefits for management.

OpenShift exists both as a managed and self-managed solution on virtually any on-premises and cloud environment. In addition, **Red Hat OpenShift service on AWS** (**ROSA**) and Microsoft **Azure Red Hat OpenShift** (**ARO**) are managed OpenShift services provided by AWS and Azure.

Visit `https://developers.redhat.com/products/openshift/overview` for more information.

AWS EKS

AWS Elastic Kubernetes Service (AWS EKS) is a container orchestration solution offered by **Amazon Web Services (AWS)** that simplifies the deployment, management, and scaling of Kubernetes clusters on AWS infrastructure without the need for users to manage the underlying infrastructure.

By leveraging familiar tools and APIs, users can create and deploy Kubernetes clusters quickly with EKS. AWS handles the operational aspects of managing the Kubernetes control plane and worker nodes, allowing users to concentrate on their applications and services rather than the underlying infrastructure.

AWS EKS seamlessly integrates with various AWS services such as Elastic Load Balancing, Elastic Block Store, and Identity and Access Management to deliver a fully managed and scalable Kubernetes environment. It also offers high availability and disaster recovery capabilities with support for multiple Regions and Availability Zones.

In summary, AWS EKS simplifies the deployment and management of containerized applications at scale on AWS infrastructure while providing an extensive range of features and capabilities.

Azure Kubernetes Service (AKS)

AKS is a managed container orchestration service provided by Microsoft Azure. It allows users to deploy and manage Kubernetes clusters on Azure infrastructure without the need to manage the underlying infrastructure themselves.

With AKS, users can quickly create and manage Kubernetes clusters using familiar tools and APIs, while Azure takes care of the operational aspects of managing the Kubernetes control plane and worker nodes. This simplifies the deployment and management of containerized applications, as users can focus on their applications and services rather than the underlying infrastructure.

AKS integrates with other Azure services, such as Azure Active Directory, Azure Virtual Networks, and Azure Storage, to provide a fully managed and scalable Kubernetes environment. It also supports multiple Regions and Availability Zones, providing high availability and disaster recovery capabilities.

In addition, AKS provides a range of features for managing and monitoring Kubernetes clusters, such as automatic scaling, rolling updates, and integration with Azure Monitor for logging and monitoring.

Overall, AKS makes it easy for users to deploy and manage containerized applications at scale on Azure infrastructure, while reducing the operational overhead associated with managing Kubernetes clusters.

VMware Tanzu Kubernetes Grid (TKG)

VMware TKG is a robust and scalable Kubernetes solution designed for enterprises. TKG empowers organizations to effortlessly deploy, operate, and govern Kubernetes clusters across diverse cloud environments, such as on-premises data centers, public clouds, and edge locations. With TKG, enterprises can seamlessly manage their Kubernetes deployments across multiple infrastructures, enabling a consistent and unified experience regardless of the underlying cloud environment.

With TKG, users can quickly and easily deploy production-ready Kubernetes clusters using a standardized, consistent architecture across multiple environments. This enables teams to focus on developing and deploying containerized applications, rather than managing the underlying infrastructure.

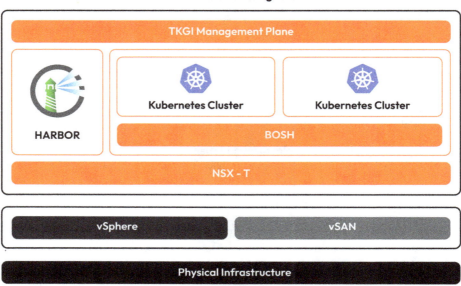

Figure 2.8 – Components of TKG

TKG integrates with other VMware solutions, such as vSphere and NSX-T, to provide a seamless and consistent experience for deploying and managing Kubernetes clusters. It also supports a range of Kubernetes distributions, including open source Kubernetes, as well as commercial Kubernetes distributions such as VMware Tanzu Kubernetes.

In addition, TKG provides a range of features for managing and monitoring Kubernetes clusters, such as automated updates, security compliance, and integration with popular monitoring and logging tools.

Overall, VMware TKG provides an enterprise-grade Kubernetes solution that enables organizations to deploy and manage Kubernetes clusters across multiple cloud environments, while simplifying the operational aspects of managing Kubernetes clusters.

HashiCorp Nomad

HashiCorp has a product that supports containers, and in some ways it works similarly to how Kubernetes works and manages applications. But in addition to containers, it also supports non-container workloads.

With its great integration with other HashiCorp products, such as Consul, Vault, and Terraform, application management and orchestration become easy.

Getting started with Nomad is pretty straightforward, and you can follow the tutorials at `https://www.nomadproject.io/`.

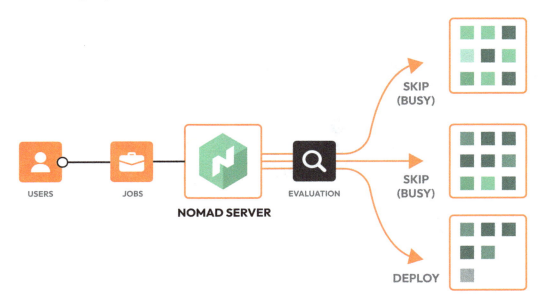

Figure 2.9 – HashiCorp's Nomad server

The preceding diagram depicts a Nomad server at work and deploying workloads to nodes.

Google Kubernetes Engine (GKE)

GKE is a scalable container service provided by Google Cloud, and it comes as a managed service. It uses Kubernetes and additional functionalities, and you can use all the Kubernetes functionalities on GKE.

As shown in the following diagram, it has a control plane, worker nodes, and additional services that can be connected to it.

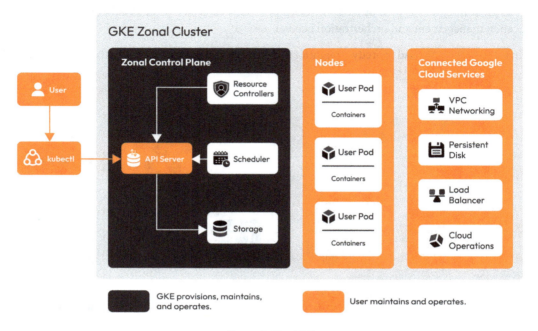

Figure 2.10 – GKE

GKE also has two kinds of nodes:

- **Control plane**: The control plane runs various processes, including the Kubernetes API server and the core resource controller. GKE manages the control plane and upgrades to new Kubernetes versions.

- **Node**: A cluster can have multiple nodes, and these nodes run the workloads. These nodes are managed by the control plane.

In most container orchestration tools, you will encounter master nodes and worker nodes, and as mentioned before, Kubernetes is the underlying technology for most container orchestration tools.

Docker Swarm

Another popular container orchestration tool is Docker Swarm, which does container life cycle management. Its responsibilities include the following:

- Deployment, availability, and provisioning of containers
- Resource allocation of containers

- Load balancing and configuration of containers
- Health monitoring of containers

The architecture components of Docker Swarm are shown in the following diagram:

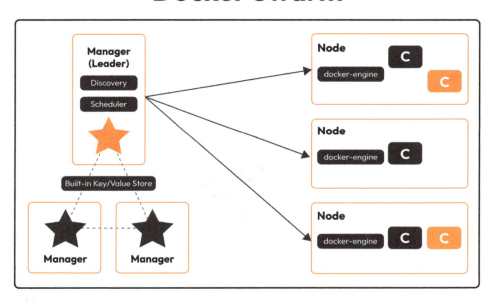

Figure 2.11 – Architecture components of Docker Swarm

The main components of Docker Swarm are as follows:

- **Docker node**: This is included in Docker Swarm and it has two kinds: master and worker
- **Manager node**: The manager node is responsible for container management and orchestration. It keeps the system in the desired state by maintaining the cluster state, scheduling services, and routing traffic to those services.
- **Docker service**: This is an executable task service.

We have discussed Kubernetes and Docker, and I want to highlight that Docker and Kubernetes do not necessarily need to be together.

Keep in mind that Docker is the industry standard for packing applications, and Kubernetes uses Docker to deploy and manage containers. You can use Kubernetes with Docker and run your applications at scale.

In this section, we looked at how containers are orchestrated. In the next sections, we will look at how do agile delivery in a standard manner across various environments and in a hybrid cloud setup.

CI/CD on the hybrid cloud

A container stack is composed of various layers that provide a clear boundary between the level of control developers and IT operations need without compromising the other team's principles.

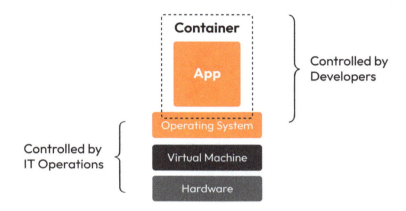

Figure 2.12 – Separation of duty between Ops and Devs

IT operations control the hardware, the virtualization layer, and the OS and make sure they are patched, secure, and managed in accordance with their best practices. Furthermore, they know how to run containers effectively on top of that infrastructure to maximize utilization.

Developers control the applications, frameworks, and libraries used in the application and package it as a container that is delivered to IT operations. Using the same tools and at the same time giving each team freedom to align their work with their requirements and goals mitigates the traditional friction between the two teams in the software delivery process.

Inside an organization, containers come to the rescue as the common language between developers and IT operations. Containers can drive a revolutionary change in how these two sides of an IT organization work together; they allow developers and the line of business to own their applications in the form of containers, which then can be deployed, maintained, and operated on top of the infrastructure that the Ops team provides, without compromising security or compliance. The promise of containers is that everybody's needs are fulfilled, and these teams can work together in better ways than they were able to previously.

Containers accelerate DevOps adoption. Before we get into how, let's clarify a few acronyms.

The two most popular acronyms in the world of cloud and DevOps are CI and CD. **CI** means **continuous integration** and **CD** is **continuous delivery**.

The diagram here shows the steps involved in CI/CD at a high level:

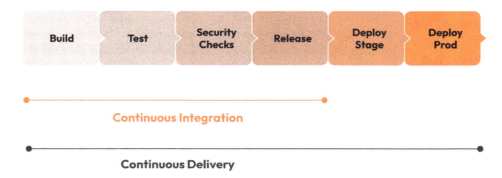

Figure 2.13 – Release pipeline

CD is also often confused with continuous deployment. Let's look at how **continuous delivery** differs from **continuous deployment** and we can then jump into continuous delivery across containers:

- **Continuous delivery**: Once you are done with CI, where code merges happen in the central repository, your next step is to deploy the changes into the production environment. Continuous delivery ensures that either on-demand or on an automatic basis, your code is getting released.

- **Continuous deployment**: Once your software is ready for deployment, you have to promote it from environment to environment to test the application in different environments. So, continuous deployment is about getting across environments. It does not involve any human intervention. You are deploying the code for end user consumption.

CI/CD has also evolved to address new packaging and deployment formats. With the introduction and adoption of technologies such as containers, Kubernetes, and Serverless, cloud-native CI/CD is becoming a preference.

Let's look at how traditional CI/CD and cloud-native CI/CD differ:

Cloud-native CI/CD	Traditional CI/CD
Designed for container applications	Designed for virtual and traditional applications
Native to Kubernetes	Not interoperable with Kubernetes resources
Less overhead	Need more resources for upkeep

Table 2.2 – Traditional versus cloud-native CI/CD

Now, as we have seen, packaging an application is simple: all you need to do is create a Dockerfile and type in your code/instructions. After that, you run a `build` command to build an image and a `run` command to run the image. Check out the *About OCI and Docker* section to refresh your memory on these commands.

Taking advantage of this simplicity, dev teams in each iteration can build a Docker image of the application once and ship that image to be deployed in various environments. The Docker image can be deployed on target environments in the exact same way regardless of whether it's a physical, virtual, or private or public cloud infrastructure.

The container packaging decouples the application from the infrastructure and enables portability across the delivery cycle and infrastructure.

What does a deployment pipeline look like with containers?

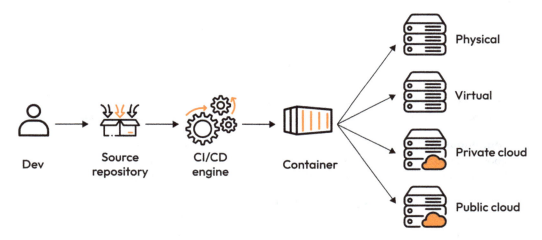

Figure 2.14 – Deployment pipeline

The process commences when a developer commits a change to the source repository. Upon notification, the CI/CD engine (e.g., Jenkins) retrieves the application code from the source repository and initiates the rebuilding of the application incorporating the new change. With the introduction of containers, the CI/CD engine takes the further step of packaging the application as a container image.

This container image can be deployed on various target environments, such as physical or virtual infrastructure, as well as private or public cloud environments. Irrespective of the underlying infrastructure, the containerized application runs consistently and uniformly, ensuring a seamless experience across different deployment environments.

For hybrid clouds, CI/CD plays an important role. Using the same methods and processes, development teams can deploy and promote software and changes in complex and heterogeneous environments running on-premises and on public clouds.

Organizations also look to cloud-native technologies to build applications that provide consistency in development across different clouds.

This is the CNCF official definition of cloud-native v1.0 (`https://github.com/cncf/toc/blob/main/DEFINITION.md`):

> *Cloud native technologies empower organizations to build and run scalable applications in modern, dynamic environments such as public, private, and hybrid clouds. Containers, service meshes, microservices, immutable infrastructure, and declarative APIs exemplify this approach.*
>
> *These techniques enable loosely coupled systems that are resilient, manageable, and observable. Combined with robust automation, they allow engineers to make high-impact changes frequently and predictably with minimal toil.*

We can use the cattle versus pet analogy; cloud-native applications are cattle. You can easily replace these applications instead of troubleshooting them.

The attributes of cloud-native technology address many challenges when adopting different clouds and infrastructures to implement a hybrid cloud.

Cloud-native applications use microservice design principles and containerization.

The tool and technology stack used to support cloud-native includes containers, Kubernetes, and DevOps.

Containers help with cloud-native in the following ways:

- **Portability**: Containers allow you to move your application code from one environment to another – development to testing to production – and provide portability.

- **Consistent environment**: *But it works in my environment* is one of the most common lines we hear when there are issues in production systems. Application teams can use containers and be sure that the code written by them will run the same anywhere they deploy and run containers.

- **Control**: Since your application code is dependent on certain libraries and dependencies, containers enable you to decide the content of your container image, giving you full control.

In this section, we learned about the important role of CI/CD in software and the hybrid cloud. We also learned about cloud-native CI/CD. You can find out how tools such as Jenkins and Tekton help implement cloud-native CI/CD.

Summary

The hybrid cloud model is going to be around for some time, and it helps with scaling on-premises architecture and enables the optimum use of public cloud offerings. Creating a hybrid cloud needs various tools, and you also need to account for existing processes and technologies that are in production.

In this chapter, we learned about hypervisors, VMs, containers, and CI/CD and how they play vital roles in cloud and hybrid cloud architectures. Making them all work together requires new processes and tooling, and using automated CI/CD that handles testing the code and deploying it to the right destination frees up your developers and infrastructure teams from time-consuming operational tasks.

In the next chapter, we will look at provisioning infrastructure with **Infrastructure as code (IaC)**.

Further reading

- Learn about Docker: `https://docs.docker.com/get-docker/`
- Learn about Nomad: `https://www.nomadproject.io/`
- Learn about AWS EKS: `https://aws.amazon.com/eks/`
- Learn about Azure OpenShift: `https://www.redhat.com/en/technologies/cloud-computing/openshift/azure`
- Learn about Buildah: `https://buildah.io/`
- Learn about Podman: `https://podman.io/`
- Learn about VMware Tanzu: `https://tanzu.vmware.com/`
- Apache Mesos: `https://mesos.apache.org/`
- Cloud-native CI/CD by Jenkins: `https://jenkins-x.io/`
- Cloud-native CI/CD with Tekton and ArgoCD on AWS: `https://aws.amazon.com/blogs/containers/cloud-native-ci-cd-with-tekton-and-argocd-on-aws/`

3
Provisioning Infrastructure with IaC

In the previous chapters, you learned about containers and VMs, the building blocks of hybrid cloud infrastructure. This chapter will show you how to use the hybrid cloud by explaining the methodologies needed to provision and operate a hybrid cloud infrastructure. We'll look at the evolution of infrastructure provisioning over the years, from ticket-based request systems to software-defined, automated, self-service systems – this transformation from purpose-built systems configured by operations specialists to software-abstracted systems that follow agile software development methodologies. We'll also explore the tools you can use to configure and deploy the infrastructure and needed services.

This chapter will cover the following topics:

- Infrastructure provisioning overview
- Virtualizing hardware with **Software-Defined Infrastructure (SDI)**
- Provisioning and managing infrastructure using **Infrastructure as Code (IaC)**
- Accelerating IT service delivery with DevOps
- Automating delivery and deployment with GitOps

Infrastructure provisioning overview

Every time we power on a PC or phone, we see it automatically initiate certain applications or processes before it's available for use. Over time, as new versions of OSs or apps become available, they can be updated by restarting the app or device. If we need to install a new app, we need to follow the process of acquiring, configuring, and installing it.

An IT department in an enterprise needs to follow a similar process to provision the requested resources, except that the numbers of requested resources are in the hundreds or thousands. Before any application can be deployed, the underlying infrastructure needs to be provisioned with pre-requisite components such as the compute, memory, OS, and runtime environment with required libraries.

Operations teams are responsible for getting the infrastructure ready for deployment and maintaining it during its life cycle. They also need to configure a technology stack consisting of VMs, containers, the network, storage, and so on.

The operations team depends on sys admins, who are domain experts that not only provision the requested resources but also troubleshoot any resulting issues to keep the system running. This is a rare and expensive skill that requires significant prior experience in their domain. Also, these skills are somewhat non-transferable – that is, a sys admin will be an expert not only in a particular OS, such as Linux, but in a particular distribution, such as **Red Hat Enterprise Linux (RHEL)**, Ubuntu, or SUSE Linux. It is like a black box for anyone other than the sys admin to understand the various config options enabled – whether encryption was enabled, what versions were used for different packages, what ports have been opened up, and with what privileges.

To deal with a growing number of requests, the sys admin typically automates the provisioning process. Usually, this means creating a script with a sequence of commands that are executed at the time of provisioning or configuring a resource. This is an imperative approach where a recipe with an ordered list of commands automates the process of provisioning. This provisioning method worked for a long time as the number of resources provisioned was limited to the capacity of a private data center and the resources would have a life cycle of months or even years.

Here are some of the commonly used methods to provision or update the infrastructure:

- **PowerShell**: The **Command-Line Interface (CLI)** and scripting language used to automate the management of the Windows operating system and applications.
- **Shell/bash scripts**: Shell scripts are lists of commands to automate repetitive tasks for a Linux system and are executed in a bash shell. Sys admins use a number of different scripts to automate the task of provisioning web servers, switching over to a backup system, and so on.
- **GUI/CLI**: A **Graphical User Interface (GUI)** makes use of graphical icons such as menus to aid interaction with the underlying system. A CLI allows text-based interactions with the system through a console or terminal.

Here is an example of a Shell script to update the Fedora operating system and install the development tools and libraries:

```
#!/bin/bash
#
# Install development packages

dnf update -y
dnf groupinstall "Development Tools" "Development Libraries" -y

echo  "Your system is updated"
```

The typical process of provisioning infrastructure involves a developer (user) creating a ticket to request infrastructure (a web server to run their application), which needs to be approved by their manager and then routed to the operations team. The operations team then logs this ticket in their request system database. Eventually, this ticket is assigned to a sys admin who uses the admin console to set up the resource with the requested characteristics (a server with appropriate compute/memory, networking, storage, OS, runtime libraries, and so on). Once the request is fulfilled, the developer is informed.

As you can see in the following diagram, the process requires fairly heavy manual intervention at each stage, which adds to the time and cost.

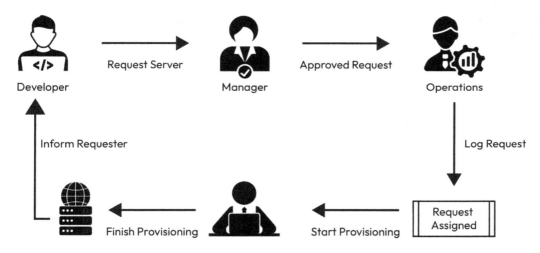

Figure 3.1 – Workflow for a ticket-based provisioning system

The frequency of application releases and related infrastructure provisioning has increased, while the duration these resources are needed for has decreased. As a result, the traditional provisioning processes are no longer sustainable.

Reducing the time needed for infrastructure provisioning and configuration has become an important goal. Organizations have also been looking for ways to reduce IT expenses by improving server utilization, gain agility, and reduce IT complexity. Virtualization and SDI are some of the main approaches used to reduce IT expenses. The adoption of cloud and hybrid technologies and solutions, beyond a pure infrastructure, has led to the adoption of IaC to standardize the provisioning and configuration across different hybrid cloud environments.

Virtualizing hardware with SDI

Virtualization popularized the concept of physical hardware (compute, network, and storage) being abstracted as software or **SDI**. This allowed businesses to improve resource utilization and flexibility while reducing provisioning time and costs. By decoupling hardware from the operating system, virtualization can provide access to abstracted resources through VMs. Refer to the description of VMs in *Chapter 1*.

The operations team uses a management console for configuring and provisioning the virtualized infrastructure requested by the development teams. Virtualization allows for policy-based infrastructure provisioning and automation. The management dashboard facilitates real-time monitoring of this SDI. The dashboard allows the VM's CPU, memory, storage, network, and so on to be configured.

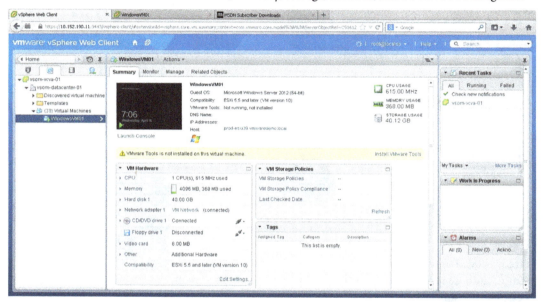

Figure 3.2 – VMware Fusion VM console for configuring CPU resources

OpenStack pools a large number of resources to deliver **Infrastructure as a Service (IaaS)**. These resources could be bare metal, VMs, or containers. *Figure 3.3* shows OpenStack, an open source cloud computing platform deployed as IaaS.

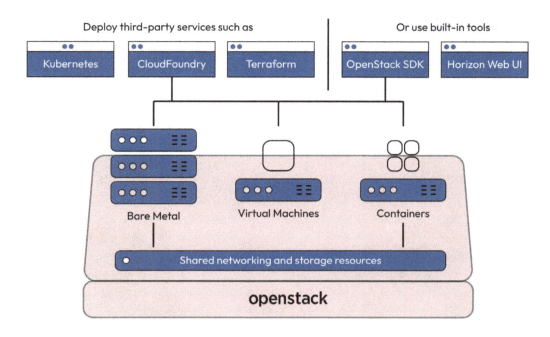

Figure 3.3 – OpenStack provides IaaS

Although SDI increased the resource utilization of hardware, the provisioning process still required the developer (user) to create a ticket to request infrastructure, which was then provisioned by the operations team. This process still required manual steps, adding to the time and cost.

Provisioning IaaS

With the advent of the cloud, infrastructure resources are now already available and can easily be requested by anyone with a credit card. Developers can easily provision the infrastructure resources themselves, test the application, and then tear down the resources. The cloud has now become a key part of infrastructure, along with on-premises infrastructure.

For instance, AWS **Elastic Compute Cloud** (**EC2**) allows you to create virtual servers, called instances, with the needed resources and security configurations. Developers can launch instances instantly with the needed amount of CPU, storage, and network resources. The EC2 dashboard provides the status of the instances in use and allows you to launch additional instances. The left-hand pane lists other services, access controls, security, and so on.

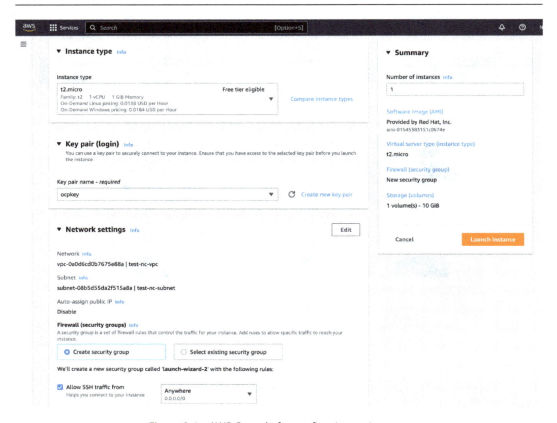

Figure 3.4 – AWS Console for configuring an instance

The EC2 dashboard contains resources and the ability to launch an instance. On the left-hand side of the dashboard, there are many links, including for EC2 limits, instances, AMIs, security groups, and **Secure Shell (SSH)** keys.

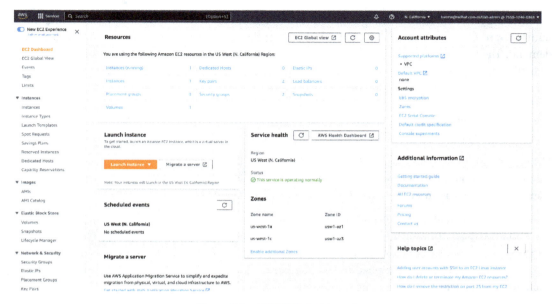

Figure 3.5 – AWS EC2 dashboard

The CLI can also be used to launch, stop, or list the cloud resources:

```
ishuverma@MacBook-Pro / % aws ec2 describe-instances
{
    "Reservations": [
        {
            "Groups": [],
            "Instances": [
                {
                    "AmiLaunchIndex": 0,
                    "ImageId": "ami-01545383151c0674e",
                    "InstanceId": "i-0c431fc7d3f322984",
                    "InstanceType": "t2.micro",
                    "LaunchTime": "2023-01-23T00:39:19+00:00",
                    "Monitoring": {
                        "State": "disabled"
                    },
                    "Placement": {
                        "AvailabilityZone": "us-west-1a",
                        "GroupName": "",
                        "Tenancy": "default"
                    },
                    "PrivateDnsName": "ip-10-0-0-140.us-west-1.compute.internal",
                    "PrivateIpAddress": "10.0.0.140",
                    "ProductCodes": [],
                    "PublicDnsName": "",
                    "State": {
                        "Code": 16,
                        "Name": "running"
```

Figure 3.6 – The AWS CLI for listing the status of current instances

Another aspect of cloud computing is the scale of available resources. Unlike the private cloud, the number of resources available is almost infinite, so workloads can be scaled up/down quickly and on demand. The life cycle of these resources is also drastically short, making it economically infeasible or inefficient to provision these resources manually. Unlike a private data center where the company has the upfront costs of owning and operating the infrastructure, whether it's used or not, the pay-as-you-consume model of cloud computing incentivizes companies to spin up resources only when needed. Operations teams need a different way of provisioning this dynamic infrastructure, one that is autonomous and doesn't require human intervention.

Using a GUI/CLI for provisioning is easy but it only works for a single developer or a small team using a limited number of resources. As the number of resources increases to more than hundreds, standardized APIs are used to create, expose, and manage cloud resources.

IT teams are faced with the challenge of provisioning virtual environments with each provider having their own method of provisioning and managing their resources through a unique GUI or APIs. As the number of services offered by each cloud provider continues to increase exponentially (in the hundreds), and this is multiplied by the number of cloud providers, each with its own API, the challenges for IT teams have become significant.

To support the needs of various stakeholders in this technically complex environment, operations teams need a simpler yet extensible way to handle infrastructure across the hybrid cloud. IaC is one such approach to managing a diverse hybrid cloud infrastructure.

Provisioning and managing infrastructure with IaC

IaC takes the SDI approach to the next level by managing IT infrastructure using code instead of manual processes or interactive configuration tools. With IaC, you can design, build, deploy, and manage the hybrid cloud infrastructure as code. The desired infrastructure specifications are defined in a configuration file, which is stored in a version control system. This specification is then implemented by an orchestrator, who provisions the desired resource. IaC enables operations teams to automate the provisioning and management of the infrastructure rather than configuring it manually.

Operations teams face significant challenges when managing various hybrid cloud environments – for example, Kubernetes clusters spread across different cloud providers. They also need to frequently deploy code changes (many times per day) and scale these deployments dynamically. In order to keep costs down and services up, computing resources need to be highly elastic and responsive to frequent changes. This complex infrastructure environment can't be configured manually as any misconfiguration could affect service availability or result in security vulnerabilities.

Once the desired infrastructure specification is defined, an orchestrator is needed to provision the desired infrastructure. Terraform, Azure Resource Manager, and Kubernetes controllers are all examples of orchestrators that provision the desired infrastructure resources in the hybrid cloud.

Here are some of the benefits of IaC:

- **Consistent**: As the same infrastructure can be deployed again and again, this reduces drift. This enables the infrastructure to be consistent across different environments such as development, test, and production.

- **Cost-effective**: Testing the desired infrastructure in software makes it more cost-effective than testing it on actual hardware.

- **Speed**: Automation allows for faster time to production as no manual intervention is needed to deploy infrastructure.

- **Stable and secure**: Automatically testing and security-scanning the infrastructure code before and after it is deployed delivers a much more stable and secure infrastructure.

- **Scalable**: Repeatable process with deterministic results from 1 instance to n instances.

- **Version-controlled**: Saving the configuration files in a version control system allows for sharing, source tracking, versioning, and change history records. The changes are auditable and can be reverted if there is an issue.

- **Developer-friendly**: Developers can provision the local dev environment with the same configuration as the production environment.

- **Documented**: Configuration is documented and available instead of being black magic known only to sys admins.

Before diving into the specific IaC provisioning tools, let's first look at the different programming approaches used by these tools.

Imperative and declarative frameworks

Imperative and declarative frameworks represent two different approaches to programming. An imperative approach describes "how" to achieve a specific task or desired end state through step-by-step instructions that not only define the sequence of operations but how they should be carried out.

A declarative approach, on the other hand, focuses on "what" or the desired outcome you want to achieve without specifying how to achieve it. In the case of infrastructure, it involves describing the desired state of resources you want to deploy and the controller taking care of the implementation details.

The following are examples of imperative and declarative frameworks:

- **Imperative frameworks**: C, C++, Java, and AngularJS
- **Declarative frameworks**: HTML, React, and SQL
- **Both**: JavaScript and Python

Here are some of the differences between these two frameworks.

Programming style

The imperative framework uses a procedural style of coding, where the programmer needs to explicitly define procedures that specify how the program should perform specific actions.

The declarative framework uses a functional style of coding, where the programmer describes what the program should achieve, using functions and expressions that specify the desired outcome. In other words, it focuses on results, not the processes.

Control flow

The computer executes code sequentially from the first line to the last line unless it encounters a construct that changes the control flow, such as a conditional statement or loop.

In the imperative framework, the programmer has fine-grained control over the control flow (the order in which the computer executes statements) of the program. Programmers need to explicitly specify the order in which instructions are executed.

As the imperative approach allows for complete control, the system could be finely customized to specific requirements. However, this benefit could turn into a disadvantage as human error can significantly increase the chances of misconfiguration. As a result, this could produce non-deterministic results, meaning the end result could be different every time based on certain conditions. This causes drift – a gap between the desired state and the end state.

In the declarative framework, the control flow is determined by the structure of the program and the order in which functions are called. The declarative framework doesn't support constructs such as loops or conditional statements.

Complexity

For new programmers, it's easier to understand imperative code by following the sequence of steps in a program. However, for complex systems with many moving parts, the imperative approach becomes more challenging for new programmers, as it requires more code to describe implementation details such as the control flow and mutable state.

The higher level of abstraction with declarative programming results in simpler and more concise code, making it much easier to get started without needing any special domain knowledge of how to achieve a specific task. A good example is DevOps, where developers benefit from the simplicity of imperative code to provision the needed infrastructure themselves instead of waiting for the operations team.

Idempotency

One of the key characteristics of the declarative approach is idempotency – no matter how many times the code is run, the result will be the same. This immutable approach means that whenever the configuration file defining the desired infrastructure is run, it will deploy the exact same infrastructure each time. Since the hidden algorithm is responsible for implementation, it limits the choice of customization or fine-tuning as compared to the imperative approach.

Debugging

Debugging and maintenance get more challenging in the imperative framework for the large code base. Debugging in the declarative framework can be easier since the code is modular and less complex. A good example is functional programming, where the output of a function should always be the same for a given input. By only producing the desired output and eliminating any side effects, it makes it much easier to test the declarative code.

Imperative and declarative framework tools for IaC

Now that we understand the differences between these frameworks, let's look at the approaches used by various IaC tools.

Here are some imperative framework tools:

- **Chef**: Chef is a popular tool for automating infrastructure provisioning using the imperative framework. It uses Ruby as the language to define the configuration.

- **AWS CLI**: The AWS CLI uses an imperative approach to manage AWS services.

Infrastructure automation tools such as Puppet, Terraform, AWS CloudFormation, and Azure Resource Manager use a declarative approach:

- **AWS CloudFormation**: This is the default tool to use to configure, provision, and manage AWS infrastructure. Resources can be created/configured using a GUI (CloudFormation Designer) or defined in a YAML/JSON file.

- **Azure Resource Manager**: The default tool used to configure, provision, and manage Azure infrastructure. Resource Manager is responsible for receiving requests from APIs and tools and provisioning the resources. The resources can be created using the web app (the Azure portal) or defined in a JSON template or Bicep (domain-specific language) file.

- **Google Cloud Deployment Manager**: The default tool used to configure, provision, and manage Google Cloud infrastructure. The resource configuration is defined in a YAML file.

- **Terraform**: A cloud-agnostic tool used for provisioning/deprovisioning cloud infrastructure (servers, networks, firewalls, and managed services). This open source tool from HashiCorp is used to build, update, and manage infrastructure as code across IT operations and developer teams. These teams can compose infrastructure as code in a configuration file using **HashiCorp Configuration Language** (**HCL**) to provision resources from any cloud provider. Terraform takes a declarative approach where the HCL file defines the end-state of infrastructure. This simple approach makes it very easy for developers to quickly spin up resources, test their apps, and spin down resources. By adding Terraform to the **Continuous Integration/Continuous Delivery** (**CI/CD**) workflow, developers can deploy infrastructure in the same pipeline as their application.

In the following example, the following configuration sets up resources for an AWS provider in the us-east-1 region. Multiple providers can be listed in the same configuration file. The resource is a c3.large instance named application_server.

After validating the configuration, the apply command is used to create the desired configuration:

```
terraform {
  required_providers {
    aws = {
      source  = "hashicorp/aws"
      version = "~> 4.16"
    }
  }

  required_version = ">= 1.3.5"
}

provider "aws" {
  region  = "us-east-1"
}

resource "aws_instance" "application_server" {
  ami           = " ami-01545383151c0674e"
  instance_type = "c3.large"

  tags = {
    Name = "ExampleAppServerInstance"
  }
}
```

Mixed approach

Some of the IaC tools use the imperative and declarative frameworks together to get the benefits of both approaches:

- **Ansible**: An open source IT automation tool used to configure, deploy software, and schedule tasks such as continuous deployments. Ansible allows both imperative and declarative approaches.

  Although Ansible is widely used to provision IaC, it provides great capabilities for configuration management too, so you can choose between applying configuration changes to already deployed infrastructure or redeploying it with the new configuration.

  Ansible Automation Platform includes multiple component libraries or *collections* (see `https://access.redhat.com/articles/3642632`) suitable for multiple IT domains, whether cloud formation, security device configuration, or VM creation, as well as others such as storage management or Windows configuration.

  There is a vast collection of community contributions published in Ansible Galaxy with more than 30k components (see `https://galaxy.ansible.com`) for your IT automation needs.

  In the following example, Ansible makes use of built-in modules to search for, download, and install Windows updates by automating the Windows Update client. It also uses the `win_command` module to run the command via PowerShell on the target host:

```
- name: Install all updates and reboot as many times as needed
  ansible.windows.win_updates:
    category_names: '*'
    reboot: yes


- name: Install all security and critical updates without a
scheduled task
  ansible.windows.win_updates:
    category_names:
      - SecurityUpdates
      - CriticalUpdates


# Search and download Windows updates
- name: Search and download Windows updates without installing
them
  ansible.windows.win_updates:
    state: downloaded
# Run command in PowerShell
- name: Execute a command in the remote shell, output going to a
file on the remote
  ansible.windows.win_shell: C:\manual-config.ps1 >> C:\log.txt
```

- **Pulumi:** This is a tool that allows configuration definition in multiple programming languages, allowing IT teams to use their existing IDEs and CI/CD for provisioning infrastructure. It's a declarative tool that uses imperative language to define the end state of the infrastructure.

Considerations for IaC

One downside of IaC scalability is that one snippet of bad code or misconfiguration could be quickly replicated to millions of instances, bringing down a whole swath of infrastructure.

For instance, AWS experienced an hours-long `us-east-1` outage that degraded many of its services:

> *"An automated activity to scale capacity of one of the AWS services hosted in the main AWS network triggered an unexpected behavior from a large number of clients inside the internal network. This resulted in a large surge of connection activity that overwhelmed the networking devices between the internal network and the main AWS network, resulting in delays for communication between these networks. These delays increased latency and errors for services communicating between these networks, resulting in even more connection attempts and retries. This led to persistent congestion and performance issues on the devices connecting the two networks."*

(Source: `https://aws.amazon.com/message/12721/`)

To avoid security vulnerabilities due to misconfiguration, security scanning tools such as checkov can be used. These scanning tools can scan systems, networks, and applications for possible security vulnerabilities.

So far, we have looked at the advancements in infrastructure provisioning for hybrid cloud environments and their benefits. However, for a bigger impact, all functional groups of an organization need to be aligned with a common goal. Companies across different industries (Amazon, Netflix, Adobe, and Fidelity) increasingly use DevOps for a competitive advantage.

Accelerating IT service delivery with DevOps

A DevOps approach requires development and operations to work together to increase the speed of software development and deployment.

As per the analyst firm Gartner:

> *"DevOps represents a change in IT culture, focusing on rapid IT service delivery through the adoption of agile, lean practices in the context of a system-oriented approach. DevOps emphasizes people (and culture), and it seeks to improve collaboration between operations and development teams. DevOps implementations utilize technology — especially automation tools that can leverage an increasingly programmable and dynamic infrastructure from a life cycle perspective."*

(Source: `https://www.gartner.com/en/information-technology/glossary/devops`)

This diverts significantly from the traditional waterfall model, where each phase of software development much be finished before the next phase can be started. With this development approach, developers work independently to build the complete application before handing it over to QA for testing and then finally to the operations teams for deployment. If any issues are found during these stages, the developers need to fix the issues before something can be deployed. This process is long and slow and requires manual intervention at each level. Here are some of the reasons why this approach is slow:

- **Siloed teams**: There are different owners for each stage of the software development life cycle with clear lines of separation of responsibilities. Developers are only responsible for application development, the QA team is responsible for testing the application, and operations is responsible for deploying the application and keeping it running.

- **Dissimilar environments**: While developers build the application on their development environments, such as their laptops, they're different from the production environment used by operations, which could elicit different app behavior.

- **Sequential**: The linearly sequential approach of this software development model makes it slow. Once developers hand off their applications for deployment, it takes several days/weeks before the operations team can proceed with the deployment as they need to find a deployment window for minimal disruption. Meanwhile, the development team has moved on to working on the next release or another project. Once the operations team gets back to the developer to fix bugs found during the deployment, the development team schedules the fixes in between other projects while the operations team waits for the updated application. This cycle repeats every time there is a code release. This adversely affects how fast an organization can bring new services or add features.

In DevOps, the developer and operations team take on common responsibilities that span the application life cycle, from development and testing to deployment and operations. This requires engineers to have skills across development and operations domains. There is frequent communication between the teams in the form of daily scrums.

Another key area of DevOps is automation of the process workflows. Once deployed, the application is closely monitored for any anomalies with quick time to resolution. The features are additive and incremental so the issues can be easily identified and the application reverted to a known good state. This results in a faster time from development to production. DevOps relies not only on the software development best practices and tools but also on a shift in culture so that the teams continuously collaborate and communicate.

Unlike traditional software release practices, DevOps requires small but frequent updates. Some organizations take this further by adopting microservice-based architecture, where a large monolithic application is broken into many independent components (microservices) that serve a single function. These microservices can be scaled up or down independently of other microservices. As each microservice is owned by a small team, the flexibility and agility of the organization are increased significantly.

Here is what a typical DevOps process looks like:

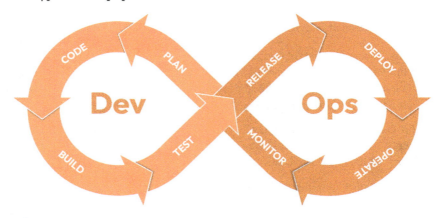

Figure 3.7 – DevOps process and areas of responsibilities

The increased release frequency leads to certain operational challenges, which are addressed by CI/CD.

Adding security to DevOps

The adoption of the hybrid cloud creates a wider attack surface than traditional on-premises deployments, which had a well-defined network perimeter. In the cloud, a small misconfiguration can expose critical information to public networks – for example, several companies, including banks and healthcare providers, leaked customer's private and sensitive (`https://krebsonsecurity.com/2023/04/many-public-salesforce-sites-are-leaking-private-data/`) information due to misconfiguration on the Salesforce Community website that allows an unauthenticated user to access records that should only be available after logging in.

DevOps teams often view security and testing as something that adds delays to their goal of releasing fast and often. However, with security gaining importance at the highest levels in companies, software needs to clear security tests before being released. By adding security later in the development cycle, the required security fixes end up costing more time and effort than if the security had been incorporated earlier in the development pipeline (shift-left).

DevOps teams often use programs, libraries, and **Software Development Kits** (**SDKs**) developed by outside vendors. This third-party code may contain security vulnerabilities at any time during the life cycle of the product. The SolarWinds software supply chain hack is a prime example of where malicious code was included during the build process of the Orion software. This software is used by thousands of companies to configure and patch their systems, putting them at great risk. See an in-depth analysis of the SolarWinds hack in this Wired article: `https://www.wired.com/story/the-untold-story-of-solarwinds-the-boldest-supply-chain-hack-ever/`.

Automated CI/CD pipelines require the use of login credentials, SSH keys, or API tokens. Careful attention must be paid to secrets management and access controls; otherwise, attackers can gain access to infrastructure and cause havoc. In one of the breaches at Uber, one of its employees uploaded AWS credentials to its GitHub repository, which allowed attackers to access sensitive rider and driver information. The source files should be scanned for credentials before they're committed to repositories (even private ones).

Vulnerability scanning should be done throughout the software development life cycle to identify and fix vulnerabilities. The **Security Content Automation Protocol** (**SCAP**) provides a set of standards used to automate vulnerability assessment, vulnerability management, and policy compliance. This specification is maintained by the **National Institute of Standards and Technology** (**NIST**). OpenSCAP is a free implementation of SCAP for Linux systems: `https://www.open-scap.org/features/standards/`.

What is DevSecOps?

Development Security and Operations (**DevSecOps**) integrates security practices at each stage of the software development life cycle rather than as an afterthought. Similar to the DevOps methodology, DevSecOps is more than the use of technology and tools, and requires a cultural change where security is a shared responsibility across development, operations, and security teams.

DevSecOps automates security testing, vulnerability scanning, and configuration management and integrates security into the CI/CD pipeline. By continuously monitoring the software supply chain and production environments, security incidents can be detected and resolved in real time.

CI/CD

CI/CD is a key part of DevOps. This iterative approach to software development comes from Agile methodology that envisions rapid code releases with a constant feedback loop between development, planning, and the customer base. By incorporating changes continuously, the problems can be discovered and remedied in real time. With all the developers in the team merging their code in a shared repository, integration issues are identified quickly through automated testing to ensure the changes haven't broken the app. It also allows all the developers to stay in sync by working on the same shared code, which incorporates other teams' updates. This approach to breaking work into smaller pieces to deliver incremental value contrasts with the waterfall model, where each developer works on their own piece of code for weeks or months before releasing it for integration. The 12 principles of the Agile Manifesto describe this approach (`https://agilemanifesto.org/principles.html`).

Continuous deployment enables software to be released to production in an automated manner. Continuous deployment can support practices such as feature flags or A/B testing, where two versions of an app with different feature sets can be compared to see which variation is preferred by the users. By gradually deploying to production and validating the software changes from various stages, any potential impact on users is limited.

Here is the workflow with CD.

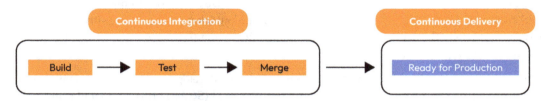

Figure 3.8 – Workflow with CD

Some of the popular CI tools are Jenkins, OpenShift Pipelines, GitLab CI/CD, GitHub Actions, Circle CI, AWS CodePipeline, and Azure Pipelines.

Refer to the section on CI/CD in the hybrid cloud in *Chapter 2* for more information on the differences for workflows with continuous delivery and continuous deployment.

Continuous testing

Unlike the waterfall model, where testing is a QA function, testing in DevOps is the shared responsibility of the whole team. Instead of a weeks-long testing cycle starting after developers have handed off code to the QA team, developers themselves test their code continuously with automated builds and unit tests so the defects are caught early. The QA team works closely with developers and with access to the most current code base to ensure software quality during each stage. Operations also works with the QA team to ensure the test environments are configured similarly to the production environment so the application behavior is the same during testing and production.

With continuous testing, the entire test process is automated to reduce the software delivery time.

Some of the common continuous testing tools are Selenium, Rational Functional Tester, and Cucumber.

Continuous operations

Continuous operations aim for 24/7 IT service availability by reducing any planned downtime such as scheduled maintenance. If any anomalies are discovered, the infrastructure is quickly returned to its desired state. All of this is accomplished through automated infrastructure provisioning and monitoring.

Monitoring and observability

Despite continuous testing, the large number of releases in DevOps means there is always a chance of an issue showing up production. This means the system behavior needs to be continuously monitored and observed to identify and fix any anomalies in the infrastructure, applications, or services. There needs to be an automatic notification process for handling any issues impacting uptime, speed, or functionality so that they can be resolved without affecting the service.

Some of the common tools used for monitoring and observability are Prometheus, Nagios, Splunk, and DataDog.

It's evident that DevOps bring many benefits to the software development life cycle. Adopting a similar approach to infrastructure operations can bring similar benefits for organizations. This is where GitOps comes into play.

Automating delivery and deployment with GitOps

Simply put, GitOps is a methodology for automating the life cycle management of applications and infrastructure. It brings the best practices of DevOps to infrastructure operations. GitOps is often used in cloud-native environments, making it a good choice for a variety of infrastructure and application deployment platforms such as Kubernetes. A case in point is Kubernetes controllers, which are responsible for making sure that the current state of the cluster matches the desired state defined in the manifest file.

Key characteristics of GitOps include the following:

- **Version control**: All of the code (the desired state of applications and infrastructure) and configuration are stored in a version control system such as Git. A Git repository becomes the **single source of truth** – the master record for the desired state of the IaC. Using a Git repository as the source of truth for configuration is the main GitOps characteristic.

- **Change mechanism**: Any changes to the code or configuration files in the repository are done using **Pull Requests (PRs)** or **Merge Requests (MRs)**. Anyone in the team can create PRs, make changes in a separate branch, test the changes, get them reviewed/approved, and make an MR to push the changes to the main branch where they can be deployed using a CD pipeline.

- **Declarative**: GitOps uses the declarative approach and is therefore well aligned with IaC principles. Any desired changes to the infrastructure can only be done by modifying the corresponding configuration files in the Git repository.

- **Automation**: Automated pipelines and tools can monitor for any changes in the Git repository and can automatically apply these to the target infrastructure. The reduced need for manual intervention eliminates the risk of human error.

Here are some benefits of GitOps:

- **Auditable and secure**: The Git repository keeps track of all changes over time so it's easy to trace who changed what over time. This Git commit history makes it easy to review, audit, and reverse changes. This also reduces the need to grant admin access to make changes to infrastructure directly.

- **Reproducible**: By not making any changes to the desired infrastructure outside of what's defined in the Git repo, the same infrastructure can be generated every single time. This also helps eliminate drift between development, test, and production environments. This can include both the platform configuration and the application configuration.

- **Faster deployments**: GitOps enables small and frequent deployments of applications and infrastructure in line with CI/CD principles. This not only results in faster deployments but also quick rollback/recovery when needed.

Here is an example of a GitOps workflow:

1. Define the desired infrastructure state in a configuration file.
2. Store the configuration file in a Git repository, which is shared between various teams.
3. Finalize and commit any changes to the configuration file.
4. Initiate the deployment of the configuration file through the CI/CD pipeline (push or pull deployment) for testing.
5. Once the deployment tests are successful, trigger the deployment to production.
6. Monitor the deployment (target environment) and configuration file (source of truth) continuously to detect any changes.
7. If any changes are detected, repeat *steps 1-5* to reconcile the difference.

Here is the visual view of the GitOps workflow:

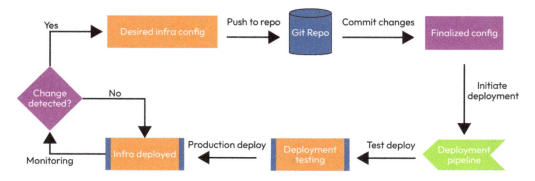

Figure 3.9 – GitOps workflow

Push versus pull deployments

Although the end goal remains the same, the desired infrastructure defined in the Git repo matches the actual deployment, there are two approaches to how the changes in the repository are applied:

- **Push deployment**: Any time there is an update to the Git repository, it triggers a CI/CD pipeline to push the update to the target infrastructure. This workflow is how most CI/CD pipelines work – for example, Jenkins, GitLab CI/CD, and Tekton.

  Here is the visual view of push deployment:

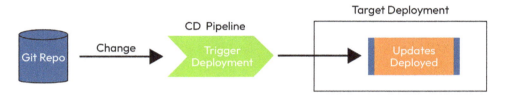

Figure 3.10 – GitOps push deployment workflow

- **Pull deployment**: A software agent in the deployed environment continuously monitors the desired state in the Git repository against the actual state of the deployment. When a change is detected in the Git repository (e.g., a new commit), the update is pulled from the Git repository and applied to the deployed environment. This workflow is used with FluxCD and ArgoCD.

  Here is the visual view of a pull deployment:

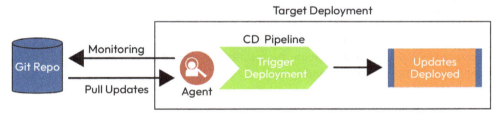

Figure 3.11 – GitOps pull deployment workflow

Let's look at an example of how ArgoCD enables GitOps.

Enabling GitOps using Argo CD

ArgoCD is a CD tool for Kubernetes that enables clusters and applications to maintain the desired state as declared in configuration files. The cluster and application configuration files are versioned in a Git repository. With ArgoCD, the entire process of application deployment and its life cycle management can be automated. ArgoCD acts as an extension of Kubernetes and helps maintain the declared state specified in a Git repository.

ArgoCD can track changes to Git repositories (or specific branches, tags, and versions) and synchronize them with the deployments in the target environment. Implemented as a Kubernetes controller, ArgoCD continually monitors the live state of applications against the desired state in the Git repo. If the live state of an application deviates from the desired state, it becomes out-of-sync and is visualized as such. ArgoCD then automatically synchronizes the deployment to its desired state.

A central Git repository with the configurations of multiple Kubernetes clusters and their applications can be used by ArgoCD to manage clusters and applications at scale as well as help with disaster recovery.

Deploying ArgoCD

ArgoCD can be deployed on Kubernetes with the following:

- The Kubectl CLI, for a manual installation
- A Helm chart for automated installation
- Operator for automated installation, upgrades, backups, monitoring, and so on

Creating an application

We need to specify information about the application and Git repository that will be monitored by ArgoCD. Add the following information to the configuration file:

- The application details (name, project, and sync policy)
- The Git repository (location and branch)
- The destination (cluster location and namespace)

A Kubernetes manifest file describes how to create and manage resources in a cluster. ArogoCD can support many methods of specifying Kubernetes manifests, including a GUI, YAML and JSON files, Helm charts, Kustomize applications, and so on.

ArgoCD will apply these configuration files to the configured clusters and applications to keep the deployment and configuration in sync.

Here is an example of a YAML file specifying application configuration:

```
apiVersion: argoproj.io/v1alpha1
kind: Application
metadata:
  name: industrial-pattern-datacenter
  namespace: argocd
spec:
  project: default
  source:
    repoURL: https://github.com/ishuverma/industrial-edge
```

```
      path: common/clustergroup
      targetRevision: main
    helm:
      valueFiles:
        - /values-global.yaml
        - /values-datacenter.yaml
  destination:
    namespace: openshift-operators
    name: in-cluster
  syncPolicy:
    automated: {}https://github.com/ishuverma/industrial-edge]
```

ArgoCD dashboard

ArgoCD can also show the status of Kubernetes resources through health checks. It provides a CLI and GUI interface for managing deployments and rollbacks and showing the state of an application.

The following figure shows the status of a resource being created by a Kubernetes controller:

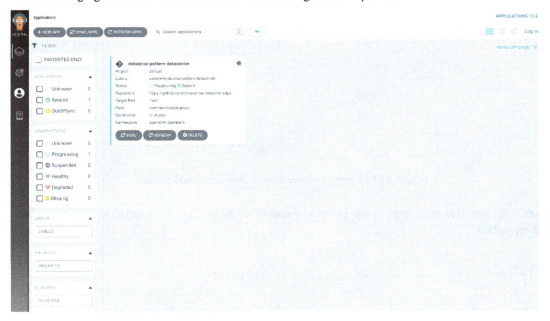

Figure 3.12 – The ArgoCD GUI console

Application monitoring and update

As ArgoCD seeks to synchronize with the Git repository, it can automatically detect and remediate any configuration drift. It continuously monitors running applications, and if their live state deviates from the desired state, it shows them as out-of-sync. The live state can then be manually or automatically synced with the desired state in the Git repository so the application remains in sync. Here is an instance of an application not being in the desired state listed in the Git repository.

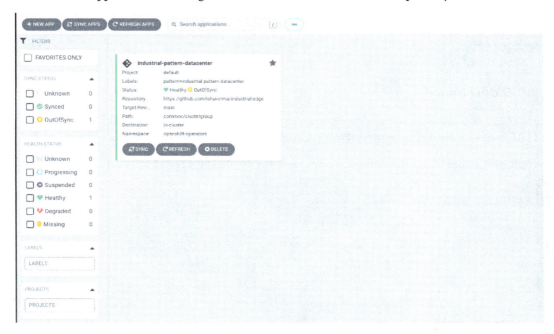

Figure 3.13 – Console showing an out-of-sync application

Here is an instance of an application called **stormshift** being synchronized with the update in the Git repository.

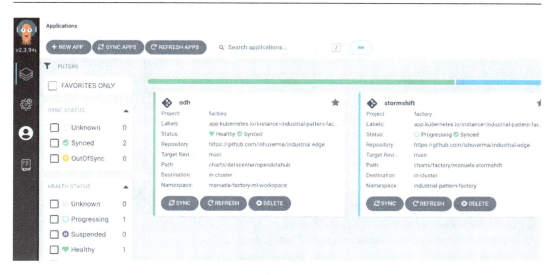

Figure 3.14 – Application syncing with the changes in the repository

ArgoCD provides a visual overview of the resulting configuration in terms of deployments, services, pods, containers, and ReplicaSets. Here is a visual representation of an application called `datacenter-gitops`:

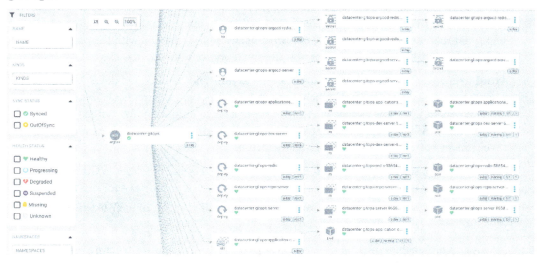

Figure 3.15 – Visual view of an application and its associated resources

ArgoCD can work with different authentication mechanisms such as OAuth2, SAML, and LDAP and enforce **role-based access control** (**RBAC**) policies. It can support various deployment capabilities including blue/green, canary, and rolling deployments, and can automatically roll back failed deployments. ArgoCD is becoming an essential tool for DevOps teams for managing and automating application deployments on Kubernetes.

Best practices for GitOps

Here are some best practices and considerations for using GitOps:

- **Separate repositories**: There should be separate Git repositories for applications and their configuration (Kubernetes manifests) and the system configuration. The application source code is totally separate from its deployment configuration. The configuration includes the desired state of deployment and includes secrets, ConfigMaps, and so on, which are independent of the application source code. Any changes to configuration should not trigger the CI pipeline for the application if there are no changes to the source code. The separation also restricts commit access to the configuration repo and prevents developers from accidentally misconfiguring the application.

- **Avoid ad hoc changes**: Sometimes, teams take a shortcut (to deal with an urgent issue or out of habit) by making direct changes to running Kubernetes clusters without going through the Git repository. These changes can cause configuration drift and adversely affect deployments as the production environment deviates from the staging environment. Tools such as ArgoCD help avoid this configuration drift by overwriting any ad hoc changes with the configuration specified in the Git repository since that is the single source of truth.

- **Avoid branches**: When deploying applications to multiple clusters, say development, test, staging, and production, separate branches are often used for each environment. This complicates the deployment workflow as each environment has its own unique configuration (secrets and ConfigMaps). A better approach is to define a core set of manifests as a base and have the environment-specific configuration as an overlay.

- **Security**: Use Git security features and RBAC to manage access and privileges to Git repositories. This ensures that only authorized users can make changes to repositories – for example, junior developers can submit PRs for updates to the application but only managers can merge the code into the master branch.

Using GitOps and IaC enables operations teams to deliver more reliable infrastructure provisioning more frequently at scale with minimal resources. As the same process deploys software applications and infrastructure, the same CI/CD pipeline could be used to deploy both. This also ensures that the infrastructure and applications are secure, auditable, and reproducible.

Summary

IaC offers many advantages over the traditional methods of provisioning infrastructure. By automating all repeatable tasks, IaC further improves the idea of SDI. Using IaC is cost-effective, fast, scalable, developer-friendly, and documented and brings consistency, stability, and security.

We learned how DevOps enables development and operations teams to work together to increase the speed of software development and deployment. It relies not only on the software development best practices and tools but also on a shift in culture where the teams continuously collaborate and communicate.

GitOps brings the best practices of DevOps to infrastructure operations. It automates the provisioning of infrastructure and deployment of applications through the use of a version control system (Git).

We also looked at the common tools used for various stages of IaC, DevOps, and GitOps.

So far, we've talked about technology concepts and frameworks in a broader sense. Now, let's dive down and look at the containerized applications themselves and how they communicate with each other in the Kubernetes world.

Further reading

- *What is Infrastructure as Code (IaC)?*: https://www.redhat.com/en/topics/automation/what-is-infrastructure-as-code-iac

- DevOps with open source: https://opensource.com/resources/devops

- GitOps with ArgoCD: https://argo-cd.readthedocs.io/en/stable/

4

Communicating across Kubernetes

This chapter is focused on exploring communication design patterns and technologies available to make independent applications work together.

A distributed solution architecture brings many independent services that are loosely coupled with each other and may have an entirely different life cycle. These services, despite the different business logic they provide, still have functions that are common and shared. These are non-functional capabilities such as logging, configuration management, debugging, security, tracing, and so on.

Application teams have a responsibility to translate business logic into working code. However, to deliver reliable applications, application teams have to spend a lot of time writing common code to deliver non-functional capabilities (as mentioned previously).

These common functions can be delivered outside of the main application logic by following a design and technical pattern known as the "sidecar" pattern. From this need for reusability, the sidecar pattern (and patterns such as the ambassador and adapter patterns) became popular and is implemented throughout the solution architecture landscape.

When deploying production-grade applications, different interfaces are used for different communication types. On one hand, you have application traffic that can come over a particular interface, but on the other hand, you have a management interface for metrics, logs, and so on. This brings the segregation of roles to the network fabric of operations.

There are four different approaches that can be used by multiple interfaces:

- **Container-to-Container** (C2C) communication
- **Pod-to-Pod** (P2P) communication
- **Pod-to-Service** (P2S) communication
- **External-to-Service** (E2S) communication

In this chapter, we will look at the following topics:

- Pod design patterns
- Container-to-container communication
- Pod-to-pod communication
- Pod-to-service communication
- External-to-service communication
- How to discover pods and services
- How to publish services
- How to stitch multiple clusters together

These patterns play a vital role in designing a hybrid cloud by allowing applications to communicate across clusters.

We will go through the preceding topics and get an understanding of networking patterns and multiple interface usage. We will also look at the aforementioned four approaches in detail and discuss service meshes.

Pod design patterns

Before we get into pod design patterns, I want to make sure that we all understand the nomenclature of Kubernetes.

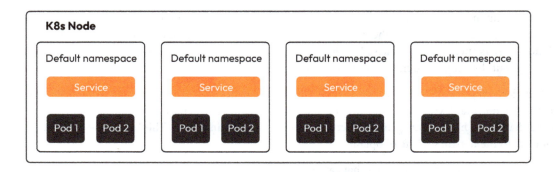

Figure 4.1 – Node, namespace, pod, and service in Kubernetes

As depicted in the diagram, we have nodes, namespaces, pods, and services:

- **Pod**: This is the smallest and simplest unit of deployment in Kubernetes, which can contain one or more containers
- **Namespace**: This is a way to create virtual clusters within a physical Kubernetes cluster to separate resources and provide access control and naming scope
- **Node**: This is a physical or virtual machine that runs containerized applications and provides the computational resources for the Kubernetes cluster
- **Service**: This is an abstraction layer that provides a stable IP address and DNS name for a set of pods and allows access to the pods from other parts of the cluster or from outside the cluster

Now, let's look at design patterns.

Kubernetes can have multiple nodes and each node can have multiple pods. These pods can have one or more containers inside them. As we know by now, a pod is a group of containers (one or more) and it is the smallest deployable unit. What you see in the following illustration is a Kubernetes cluster with multiple nodes and pods; however, the pods have only one container inside them:

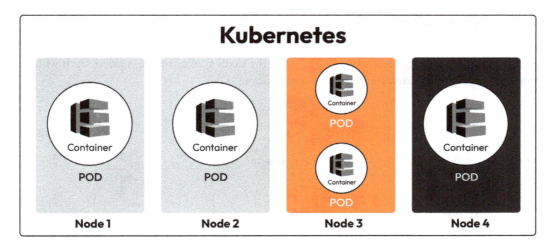

Figure 4.2 – Containers, pods, and K8s (single-container pod and a multiple-container pod)

Again, a pod can have one or more containers.

When you have containers that are closely related – the same life cycle – you may want to package them together and deploy them in the same pod. The following illustration shows a pod with multiple containers:

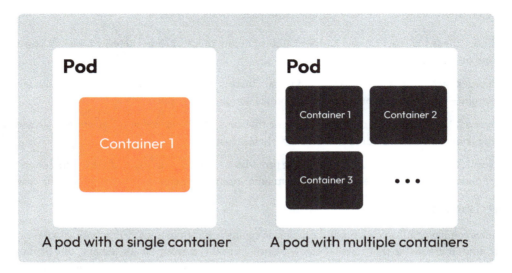

Figure 4.3 – Multiple containers in a pod

The operational burden on business application developers is reduced when containers are bundled together.

There are three multi-container pod network design patterns that involve two closely associated containers. In the following illustration, you can see the main application containers, each with a second container labeled sidecar, adapter, and ambassador:

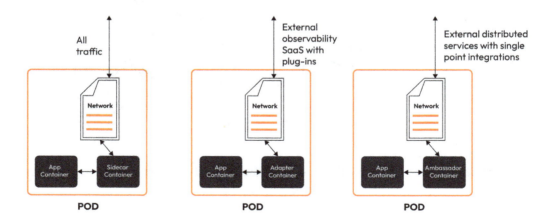

Figure 4.4 – Pod design patterns

Let's dive deep into the sidecar pattern.

The sidecar pattern

The sidecar pattern is a design pattern used in Kubernetes that involves running a container alongside a primary container in the same pod. The primary container in this case is usually the application that needs to be run, while the sidecar container provides additional functionality that complements the primary container. The sidecar pattern is implemented using sidecar containers.

The sidecar containers share resources such as network interfaces and pod storage while running alongside the main application container. A huge benefit is low latency because they communicate on the same network using localhost or a netns IP address. The functionality provided by them is non-business logic, which means that without altering application code, they can provide capabilities such as tracing, security, and so on. The sidecars don't have to be coded in the same programming language as the main application container. When you look at service mesh solutions in industry, you will see that they leverage the sidecar pattern.

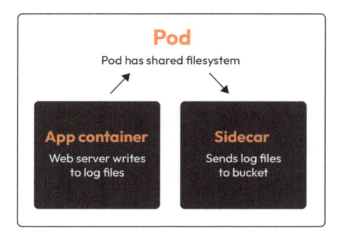

Figure 4.5 – Sidecar design pattern

Here's an example of the sidecar pattern using Kubernetes manifest files:

```
apiVersion: v1
kind: Pod
metadata:
  name: myapp-pod
spec:
  containers:
    - name: myapp
      image: myapp-image
      ports:
        - containerPort: 8080
    - name: sidecar
```

```
image: sidecar-image
ports:
  - containerPort: 9090
```

In this example, we have a pod with two containers: myapp and sidecar. The myapp container is the primary container that runs our application, and the sidecar container provides additional functionality to the application. A common use case for the sidecar pattern is to provide logging and monitoring functionality to the primary container. In this example, we can use the sidecar container to collect logs and metrics from the myapp container and forward them to a logging or monitoring system. Here's how we can implement this use case using the sidecar pattern:

```
apiVersion: v1
kind: Pod
metadata:
  name: myapp-pod
spec:
  containers:
    - name: myapp
      image: myapp-image
      ports:
        - containerPort: 8080
      volumeMounts:
        - name: logs
          mountPath: /app/logs
    - name: sidecar
      image: sidecar-image
      ports:
        - containerPort: 9090
      volumeMounts:
        - name: logs
          mountPath: /sidecar/logs
  volumes:
    - name: logs
      emptyDir: {}
```

In this modified example, we have added a volume that is shared between the myapp and sidecar containers. The myapp container writes logs to the /app/logs directory, while the sidecar container reads logs from the /sidecar/logs directory and forwards them to a logging or monitoring system.

The sidecar pattern allows us to separate the logging and monitoring functionality from the application code, which makes it easier to manage and scale the application. We can also update the sidecar container independently of the myapp container, which makes it easier to add or remove functionality without affecting the application.

In summary, the sidecar pattern in Kubernetes can be used to provide additional functionality to the primary container in a pod. In this example, we used the sidecar pattern to provide logging and monitoring functionality to the `myapp` container by running a `sidecar` container that collects logs and metrics and forwards them to a logging or monitoring system.

The adapter pattern

The adapter pattern is the same as the sidecar pattern in terms of both containers running in the same pod.

For externally pluggable observability solutions, the adapter pattern is used to offer simple integration points. For the sake of standardizing the interface between the main application container and external observability services such as logs or invoking libraries that don't have language bindings, the adapter pattern can be of great help.

In the context of **Kubernetes (K8s)** manifest files, an example of the adapter pattern could be the use of ConfigMaps to adapt the configuration of an application to be compatible with the Kubernetes API.

Let's say we have an existing application that reads its configuration from a file on disk. However, we want to deploy this application on Kubernetes and use ConfigMaps to store its configuration instead. In this case, we can use the adapter pattern to adapt the application's interface to the Kubernetes API by creating a ConfigMap that contains the application's configuration.

Here's how this use case gets implemented using the adapter pattern:

1. First, we create a ConfigMap that contains the application's configuration data. We can do this using a YAML file, like so:

   ```
   apiVersion: v1
   kind: ConfigMap
   metadata:
     name: my-app-config
   data:
     app-config.yaml: |
       key1: value1
       key2: value2
   ```

   This YAML file creates a ConfigMap with the name `my-app-config` and one key-value pair that contains the application's configuration data.

2. Next, we modify the application's code to use the Kubernetes API to retrieve its configuration from the ConfigMap instead of reading it from a file on disk. We can do this by using a Kubernetes client library such as kubectl or the Kubernetes API client.

3. Finally, we create a Kubernetes manifest file that deploys the application and mounts the ConfigMap as a volume in the container. Here's an example YAML file:

```
apiVersion: apps/v1
kind: Deployment
metadata:
  name: my-app
spec:
  replicas: 1
  selector:
    matchLabels:
      app: my-app
  template:
    metadata:
      labels:
        app: my-app
    spec:
      containers:
      - name: my-app
        image: my-app-image
        volumeMounts:
        - name: config-volume
          mountPath: /app/config
      volumes:
      - name: config-volume
        configMap:
          name: my-app-config
          items:
          - key: app-config.yaml
            path: app-config.yaml
```

This YAML file creates a deployment that deploys the application in a container. It also creates a volume that mounts the ConfigMap as a file in the container's filesystem. The `items` field in the `configMap` section specifies which key in the ConfigMap to mount and where to mount it in the container's filesystem.

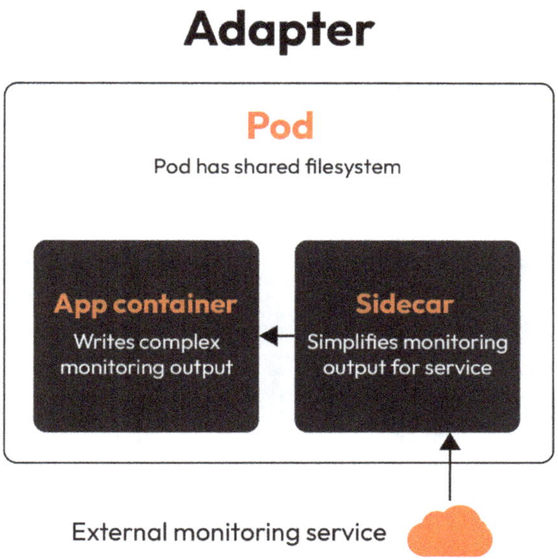

Adapter

Figure 4.6 – Adapter design pattern

By using the adapter pattern in this way, we were able to adapt the application's interface to the Kubernetes API and deploy it on Kubernetes using ConfigMaps to store its configuration.

Ultimately, the adapter takes heterogeneous containers and turns them into a consistent and unified interface for outside services.

The ambassador pattern

An ambassador container allows outside service access without implementing a service. This helps to connect the main container with the outside world, such as proxying localhost connections to the external world of the network fabric. For system integration to distributed external systems, the ambassador pattern can offer a straightforward approach.

Here's an example of how the ambassador pattern can be implemented using Kubernetes manifest files:

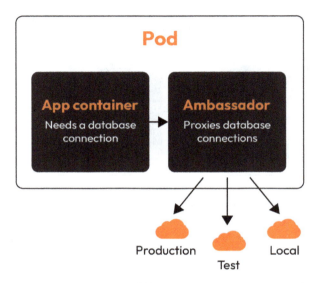

Figure 4.7 – Ambassador design pattern

Assuming we have two Kubernetes services, `service-a` and `service-b`, both of which need to be exposed to the outside world on a single IP address and port, we can create an ambassador resource as follows:

```
apiVersion: getambassador.io/v2
kind: Ambassador
metadata:
  name: my-ambassador
spec:
  config: |
    ---
    apiVersion: ambassador/v0
    kind: Mapping
    name: service-a-mapping
    prefix: /service-a/
    service: service-a
    --
apiVersion: ambassador/v0
    kind: Mapping
    name: service-b-mapping
```

```
    prefix: /service-b/
    service: service-b
```

This YAML file creates an ambassador resource named `my-ambassador` and defines two mappings that map incoming requests to the appropriate Kubernetes services. The first mapping maps requests with the prefix `/service-a/` to the `service-a` Kubernetes service. The second mapping maps requests with the prefix `/service-b/` to the `service-b` Kubernetes service.

Now, when an incoming request is received by the ambassador resource, it checks the request path and forwards it to the appropriate service based on the mapping rules. For example, if an incoming request has the path `/service-a/foo`, it will be forwarded to the `service-a` Kubernetes service.

This pattern is useful in cases where we need to expose multiple services on a single IP address and port, such as when we have a microservices architecture with many small services that need to be exposed to the outside world. By using the ambassador pattern, we can simplify the configuration and management of these services, and also provide a consistent interface to the outside world.

In a nutshell, all of these patterns offer sharing of the same pod network namespaces across containers within the pod. Because they share the same network interfaces, they are controllable by the same IP table rules in the pod network namespace.

As far as differences go, the fastest way to understand the differences between all three patterns is depicted in the following figure:

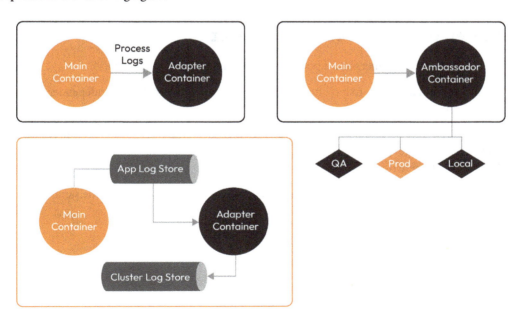

Figure 4.8 – Pod design patterns

For a quick recap, the following table contains the differences between the different patterns:

Design Pattern	Sidecar	Adapter	Ambassador
Usage	Share resources and do other functions. Sidecar containers enhance applications by providing additional functionalities such as logging, monitoring, proxying, authentication, or encryption, deployed alongside the main container in the same pod.	Check the status of other containers. The adapter pattern provides a unified interface to interact with third-party observability solutions and it can be leveraged by multiple applications.	Network/integrate with outside services. The ambassador pattern simplifies external access for apps without modifying core containers. It enables proxying, reverse proxying, request limiting, and routing, enhancing flexibility and security.

Table 4.1 – Usage of different patterns

Now that we have looked into various pod patterns, let's look into how communication across containers works.

Container-to-container communication

Pods can have multiple containers and a pod has its **network namespace (netns)**.

Network namespaces enable you to have network interfaces and IP tables that are independent of the host environment (i.e., the K8s worker node).

Containers inside a pod share the same networking namespace. Because of the shared networking namespace, containers get access to the same network resources, such as the IP address and ports through the same IP table routing logic.

Each container inside the pod can communicate via localhost as if they are part of the same netns.

As shown in the following figure, the containers inside a pod share the same netns:

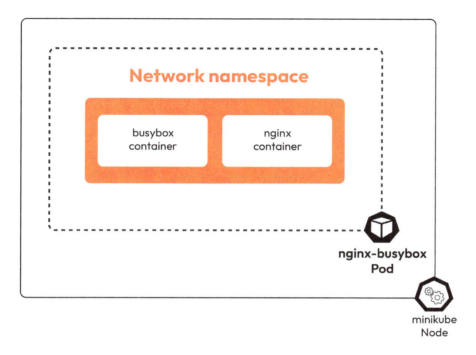

Figure 4.9 – Container-to-container communication

Having learned how containers communicate with each other, now it is time to look into how pods communicate with each other.

Pod-to-pod communication

While pods can communicate with each other using direct IP addresses, it is not the recommended way of communicating. Pods are ephemeral and are meant to be replaced, rather than managed directly.

There are a few different ways that pods can communicate with each other in Kubernetes:

- **Within the same node**: Pods can communicate with each other using localhost or their own IP addresses, just as processes on a regular host can.

- **Across nodes**: Pods can communicate with each other across nodes using their own IP addresses, just as processes on different hosts can. However, this can be difficult to set up and manage, so it is usually easier to use a service to facilitate communication between pods.

- **Using a service**: A Kubernetes service is a logical abstraction over a group of pods, and it provides a stable IP address and DNS name for those pods. Pods can communicate with each other using the DNS name of the service, rather than the IP addresses of the individual pods. This makes it easy to communicate with a group of pods, even if the pods are recreated or moved to different nodes.

In summary, the most common way for pods to communicate with each other in Kubernetes is by using services. Services provide a stable way to access a group of pods, and they can be accessed using DNS names or environment variables.

Kubernetes comes with a networking model known as a **container network interface** (CNI). In other words, Kubernetes leverages the CNI for setting up networks and, by using network plugins, its implementation can occur.

Many CNI plugins are available (Flannel, Calico, Weave Net, etc.) and you can choose depending on the need of the use case. The CNI is generally configured during the initial cluster setup and is called the master **Cluster Network Operator** (CNO) – usually the `eth0` interface in the pod.

A CNI is a standard for configuring container network interfaces, while the CNO is a Kubernetes operator that manages network-related resources and configurations in a Kubernetes cluster.

With most of the primary CNI implementations (such as `ovn-kubernetes`), the K8s node has a designed CIDR range of IP addresses for their pods. This ensures that every pod gets a unique IP address.

In many of the primary CNI implementations, such as `ovn-kubernetes`, each K8s node is assigned a specific **Classless Inter-Domain Routing** (CIDR) range of IP addresses for its pods. This allocation guarantees that every pod obtains a distinctive IP address that allows communication with other pods in the same namespace.

To enable communication between pods, network traffic needs to be transmitted between the network namespace of the pod and the network namespace of the Node/host (also known as the worker), and remain within the namespace of the tenant throughout the cluster.

To achieve this, a **virtual Ethernet** (**veth**) device or a veth pair is used to connect the Pod namespace and the host netns. These virtual pod interfaces are then linked via a virtual network bridge, which facilitates the exchange of traffic among them using the **Address Resolution Protocol** (ARP) over the project/tenant namespace. The following diagram depicts Kubernetes networking.

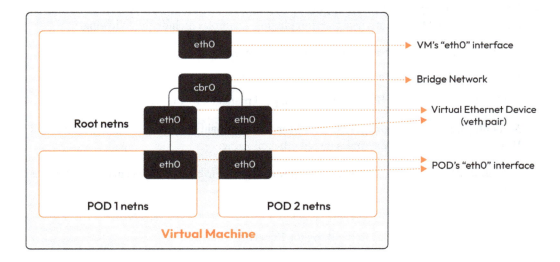

Figure 4.10 – Kubernetes network explained

In the case of additional CNI usages (for example, `macvlan`, `ipvlan`, etc.), pods can also communicate with each other with assigned IP addresses (via DHCP or static assignment) over these additional interfaces (for example, `net1`, `net2`, etc.) based on using the ARP as well.

Pods also have multiple interfaces, which provides greater flexibility and enables more complex network topologies to be implemented within a Kubernetes cluster. Let's look at pods with multiple interfaces in the next section.

Pods with multiple interfaces

CNIs provide a portable and reusable **Application Programming Interface** (**API**) to configure and use network interfaces.

There are three main categories of CNI plugins:

- **Interface creation plugins**: These plugins create and configure network interfaces for containers. Examples include the following:

  - **Bridge**: Creates a Linux bridge interface and connects it to the container's network namespace.

  - **Macvlan**: Creates a new virtual Ethernet interface with a unique MAC address and VLAN ID.

  - **Host-device**: Allows the container to use a physical network interface on the host.

- **IP address management (IPAM) plugins**: These plugins assign IP addresses to network interfaces created by the interface creation plugins. Examples include the following:

  - **DHCP**: Requests an IP address from a DHCP server.

  - **Host-local**: Assigns IP addresses from a predefined range on the host.

  - **Calico IPAM**: Assigns IP addresses from a pool managed by the Calico network plugin.

- **Meta plugins**: These plugins provide additional functionality beyond interface creation and IP address management. Examples include the following:

  - **Multus**: Allows multiple CNI plugins to be used together to create multiple network interfaces for a container.

  - **Cilium**: Implements network security policies and service mesh capabilities.

  - **Flannel**: Provides overlay networking for Kubernetes clusters.

In many cases, clusters have requirements to maintain separate interfaces for logging, monitoring, or different architecture. Thus, a separate interface is needed, and that is where Multus comes in.

Multus CNI is a plugin referred to as a meta plug-in, a CNI plug-in that can embed other CNI plug-ins. It works like an encapsulation that includes other CNI plug-ins for adding multiple network interfaces to the pod netns in the K8s cluster.

Multus CNI is the work of the Network Plumbing working group, which created the specification for `NetworkAttachmentDefinition`, a Kubernetes **custom resource definition** (**CRD**) used to express the aim and scope for which networks the pod should be plugged into. The following figure shows Multus CNI enabling multiple network interfaces on a single pod:

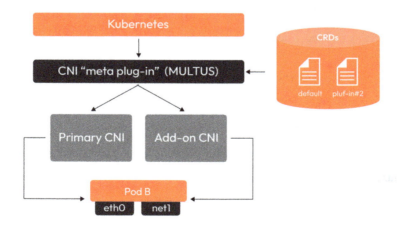

Figure 4.11 – Multus CNI

To leverage multiple networks with your containers (sitting in the same pod), first, you need to plan, design, and then implement your networking fabric in the K8s cluster. Then, you refer to those network parameters inside the K8s network definitions:

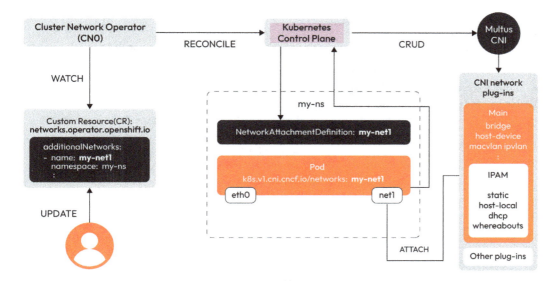

Figure 4.12 – Multus logical implementation

There are several options/solutions available for managing IP addresses on Multus add-on CNI interfaces (i.e., `NetworkAttachments`) when using Kubernetes native networking constructs. These solutions include the following:

- **Host-local**: This solution manages IP allocation within specified ranges on a single node only. The storage of IP address allocation is specified in flat files located on each host, hence it is called host-local.

- **DHCP**: This solution allocates IP addresses using an "external" DHCP server, which can be integrated with the cluster.

- **Whereabouts**: The Whereabouts pattern efficiently manages IP allocation within specified ranges across a Kubernetes cluster. It utilizes Kubernetes Custom Resources to store IP address allocation information, enabling IP allocation across any host within the cluster.

- **Static**: This solution allocates IP addresses statically for each pod, with these IPs allocated on the pod YAML or using Kubernetes configuration.

It's worth noting that these are just a few examples. The choice of solution depends on the organization's infrastructure, service-level agreement, and general requirements. Some solutions are easier to implement but might have some disadvantages, and vice versa. Now, let's consider the preceding solutions as options for an application planning to use horizontal pod autoscaling (increasing the pod count to deal with the traffic demand):

- Host-local and/or static do *not* address the need for **Horizontal Pod Autoscaling** (**HPA**) across different nodes.

- DHCP needs an external DHCP server per CIDR that a pod (or pods) attaches to, which can be costly (plus operational overhead) to deal with outside the Kubernetes cluster as these networks may stay private within the application platform network scope. Whereabouts takes an address range and assigns IP addresses within that range. Once a pod is allocated an IP address, its location is stored in a data repository by Whereabouts throughout the pod's lifespan. Subsequently, when the pod is decommissioned, the address is freed up by Whereabouts, making it available for subsequent requests. Whereabouts always assigns the first available address in the range. The following diagram depicts the Whereabouts mechanics:

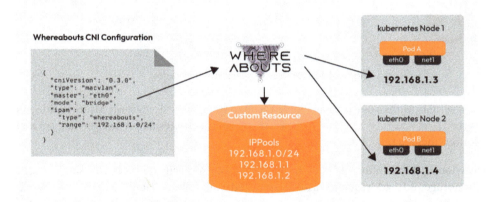

Figure 4.13 – Whereabouts mechanics

The following code defines a Kubernetes **Network Attachment Definition** (**NAD**) for a macvlan network, using the Whereabouts IPAM plugin to assign IP addresses to the pods. The NAD specifies the IP range, gateway, and exclusion list for the network, as well as the interface to be used for the macvlan connection:

```
apiVersion: "k8s.cni.cncf.io/v1"
kind: NetworkAttachmentDefinition
metadata:
  name: addl-network-1
```

```
spec:
  config: '{
      "cniVersion": "0.3.0",
      "name": "was-i2i5gcore-192.168.100",
      "type": "macvlan",
      "master": "ens224",
      "mode": "bridge",
      "ipam": {
        "type": "whereabouts",
        "range": "192.168.100.0/24",
        "gateway": "192.168.100.1",
        "exclude": [
            "192.168.100.1/32"
        ]
      }
    }'
```

This is a Kubernetes deployment manifest file. It specifies a deployment with one replica of a container named debug-app-fenar1. The container runs the latest version of the quay.io/narlabs/debugpod image and has resource requests and limits for CPU, memory, and storage. The container is annotated with the network attachment name whereabouts100 to attach it to a specific network. The deployment also specifies an image pull secret to authenticate with the container registry:

```
apiVersion: apps/v1
kind: Deployment
metadata:
  labels:
    name: debug-app-fenar1
  name: debug-app-fenar-1
spec:
  replicas: 1
  selector:
    matchLabels:
      name: debug-app-fenar1
  template:
    metadata:
      annotations:
        k8s.v1.cni.cncf.io/networks: addl-network-1
      labels:
        name: debug-app-fenar1
    spec:
      containers:
      - image: quay.io/narlabs/debugpod:latest
        name: debug-app-fenar1
        command: [ "/bin/bash", "-c", "--" ]
```

```
      args: [ "while true; do sleep 30; done;" ]
      resources:
        requests:
          cpu: 500m
          memory: 768Mi
          ephemeral-storage: 1Gi
        limits:
          cpu: 900m
          memory: 900Mi
          ephemeral-storage: 2Gi
  imagePullSecrets:
    - name: narlabs-i2i-robot-pull-secret
```

In this section, we looked into the Multus add-on CNI plugin, which provides the capability to attach multiple network interfaces to a pod.

However, managing IP addresses for these additional interfaces can be challenging. Several solutions are available for managing IP addresses on Multus interfaces, including host-local, which assigns IP addresses from a predefined range on the host; DHCP, which uses a DHCP server to assign IP addresses dynamically; static, which assigns fixed IP addresses to interfaces; and Whereabouts, which dynamically assigns IP addresses from a predefined range using a central data store. Each solution has advantages and disadvantages, and the choice will depend on the specific needs and constraints of the deployment. Lets look into pod-to-service communication.

Pod-to-service communication

Pods in Kubernetes are designed to be temporary and can be created, terminated, and scaled up or down based on the traffic demand. This means that the IP address of a pod can change frequently, making it difficult for clients to connect to them.

Kubernetes networking addresses this issue through the utilization of its service feature, which allocates a stable virtual IP address to the frontend for establishing connections with backend pods linked to the service. Additionally, the service distributes traffic directed toward this virtual IP to the group of related pods in a load-balanced manner.

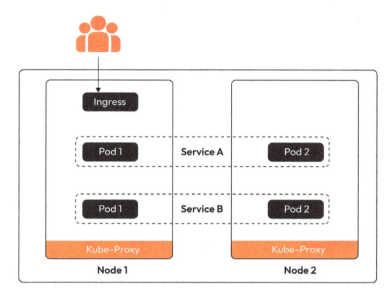

Figure 4.14 – Pods communicate via Services

Furthermore, a service tracks the IP address of a pod, so even if the pod's IP address changes, clients can still connect to it without any problem because they are only communicating with the static virtual IP address of the service. This makes it possible for clients to connect to the application without needing to know the specific IP address of the pods, reducing the amount of reconfiguration required when pods change.

Pod-to-service communication is important for service discovery and load balancing within a Kubernetes cluster. Now let's look at external-to-service communication.

External-to-service communication

In addition to routing traffic within a cluster, Kubernetes also allows you to expose an application to an external network. There are two main ways to do this:

- **Egress**: This functionality enables you to direct traffic from your Kubernetes Services to the internet. It leverages iptables to execute source NAT, thereby making it seem like the traffic originates from the node rather than the Pod.

- **Ingress**: This functionality facilitates the management of incoming traffic from external sources to Services. It also provides the ability to regulate access to Services through connection rules. Normally, two separate ingress solutions operate in distinct network stack regions: the service load balancer and the ingress controller.

The following diagram shows the ingress and egress functionality:

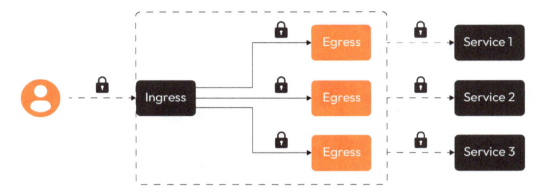

Figure 4.15 – Ingress and egress

These are the main way to expose your services to an external network, but it's worth noting that there are some other ways, such as using service meshes and gateways.

How to discover pods and services

There are several ways to discover pods and services in Kubernetes. Here are a few common methods:

- **kubectl command-line tool**: The kubectl command-line tool is the primary way to interact with a Kubernetes cluster. You can use the `kubectl get pods` and `kubectl get services` commands to list all pods and services in the current namespace. You can also use the `kubectl describe pod [pod-name]` and `kubectl describe service [service-name]` commands to get detailed information about a specific pod or service.

- **Kubernetes API**: The Kubernetes API is the underlying mechanism that kubectl uses to communicate with the cluster. You can use the Kubernetes API directly to retrieve information about pods and services. The API endpoints for pods and services are `/api/v1/pods` and `/api/v1/services`, respectively.

- **Kubernetes Dashboard**: The Kubernetes Dashboard is a web-based UI for interacting with a cluster. It provides a visual way to view and manage pods, services, and other Kubernetes objects. You can access the Dashboard by running the `kubectl proxy` command and then navigating to `http://localhost:8001/ui` in a web browser.

- **Kubernetes DNS**: Kubernetes provides a built-in DNS service for resolving pod and service names to IP addresses. You can use the DNS service to discover pods and services by resolving their names to IP addresses.

- **Monitoring solutions**: There are several Kubernetes monitoring solutions that provide an easy way to discover and monitor pods and services in your cluster. These solutions often provide visualizations, alerts, and other features that make it easy to understand the state and behavior of your applications.

It's worth noting that these are some basic methods and there are more advanced discovery methods, for example, using service meshes, External-DNS, and so on. Let's look at External-DNS.

External-DNS is a Kubernetes add-on that allows you to manage DNS records for services, ingress, and pods automatically. It can be used to discover pods and services running in a cluster by configuring DNS records that map the names of services and pods to their corresponding IP addresses.

Here are the basic steps to configure External-DNS to discover pods and services:

1. **Install External-DNS**: You will need to deploy External-DNS as a Kubernetes pod in your cluster. There are various ways to do this, but one common method is to use a Helm chart.

2. **Configure External-DNS**: Once External-DNS is running, you will need to configure it to connect to your DNS provider. External-DNS supports many popular DNS providers, including AWS Route 53, Azure DNS, and Google Cloud DNS.

3. **Create DNS records**: External-DNS will automatically create DNS records for services and pods in your cluster. By default, it will create A and AAAA records for pods and SRV records for Services. You can configure External-DNS to create other types of records as well.

4. **Verify the DNS records**: Once External-DNS is configured, you can use the DNS provider's web interface or command-line tools to verify that the DNS records have been created correctly.

5. **Test the DNS resolution**: To test the resolution of the services, you can use `nslookup` or the `dig` command on your local machine to see whether the DNS records can be resolved.

It's also worth noting that you can use External-DNS to create additional records, such as CNAMEs, to route traffic to your pods using other domain names such as `*.yourdomain.com`.

You should also keep in mind that External-DNS does not support all DNS providers, but the support is growing, so it's better to check whether your DNS provider is supported by External-DNS before using it.

We looked at the discovery of pods and services. Now, let's look at publishing services.

How to publish services

Kubernetes employs Services to enable access to a set of pods that share a common label selector. This allows communication between applications within the cluster and makes it possible to expose applications running in the cluster to external entities, as depicted in the following figure:

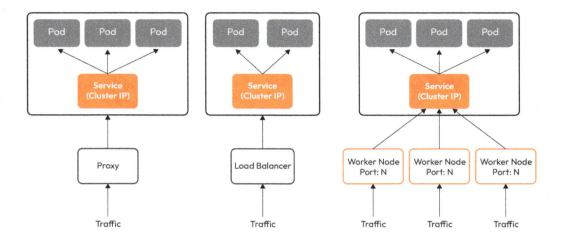

Figure 4.16 – K8s service exposure types

Kubernetes offers different service types that allow you to specify how a service should be exposed. Here are the three main types:

- **ClusterIP**: ClusterIP is the default service type that restricts access to a Service to within the cluster only. This is useful for applications that need to communicate with each other within the cluster but do not need to be accessed from outside the cluster.

- **NodePort**: By using this configuration, the Service becomes accessible from external sources outside of the cluster. To achieve this, a specific port is opened on all nodes to forward traffic to the Service. This is not recommended for production environments unless the cluster has a small footprint and you are working with limited network accessibility.

- **LoadBalancer**: This is a service type that allows the external exposure of a Service through the on-premise or cloud provider's load balancer. Traffic received by the load balancer is forwarded to the backend pods. If you are using a public cloud infrastructure, such as AWS or GCP, this will likely be an additional cost item, but it also offloads the burden of maintaining the load balancer. Some enterprise Kubernetes solutions such as OpenShift include an integrated load balancer for easy configuration and maintenance.

It's worth noting that ClusterIP is the best option for P2P communication, which avoids isolating the services from the outside world, and also from a cost perspective. The other two options will come with some additional cost – scaling costs with NodePort and SaaS costs with LoadBalancer.

Now that you know how to publish services, let's move on to the next section and see how to stitch multiple clusters together.

How to stitch multiple K8s clusters

If you have a multi-K8s cluster deployment blueprint and do not want every possible communication to traverse the internet freely, you may consider implementing a secure (encrypted) interconnect networking solution between clusters.

We are going to look at different technologies to enable multi-cluster communication. Here are the three technologies that we are going to look at:

- Submariner
- Skupper – using a common application network (layer 7)
- Service mesh

Submariner – using layer 3 networking

Submariner is a software component that enables the seamless connectivity of pods and services running in different clusters. It allows pods and services running on different clusters to communicate with each other as if they were on the same network.

Submariner consists of several different components that work together to provide this connectivity. These components are listed here:

- **Lighthouse**: This is a key component, which runs as a pod in each cluster and is responsible for discovering and maintaining the state of other Lighthouse pods running in other clusters.
- **Gateway**: Another component is the gateway, which runs on each cluster and is responsible for routing traffic between clusters.
- **Submariner-agent**: Finally, the Submariner-agent runs on each cluster and is responsible for configuring the necessary networking rules to allow for inter-cluster communication.

These Submariner components work together to route traffic and manage the connectivity between pods.

Submariner architecture

How it works is Submariner creates a VPN connection between each cluster, allowing the clusters to communicate with each other securely over the internet. The control plane creates a virtual network overlay on top of the existing network infrastructure, and the data plane forwards traffic between the clusters using this overlay. Services running in one cluster can be accessed from another cluster by resolving the service name to the appropriate IP address. Submariner also allows applications/namespaces to span multiple clusters, providing additional benefits such as high availability and disaster recovery.

The diagram here illustrates Submariner connecting pods in different clusters:

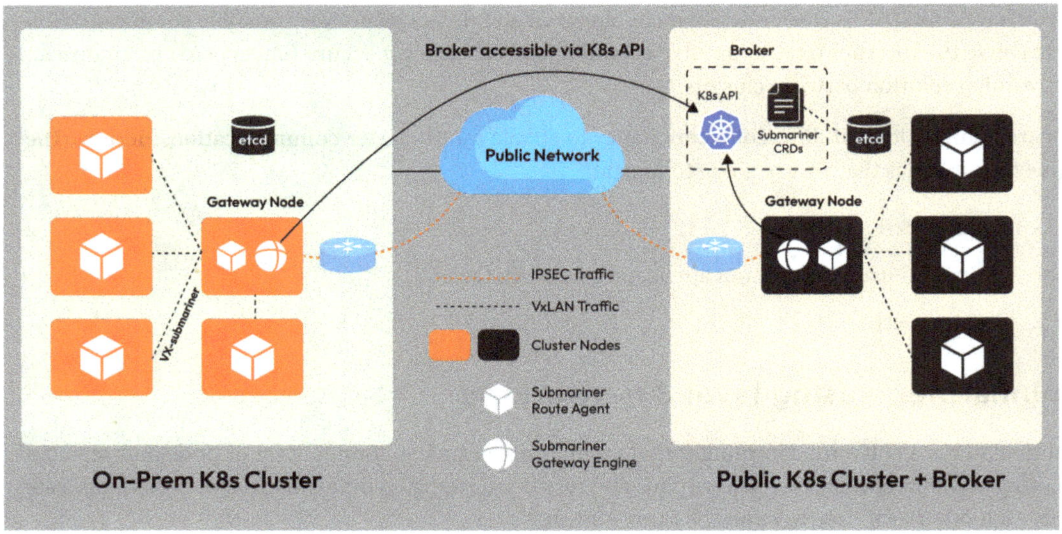

Figure 4.17 – Submariner architecture

Let's take a look at the components shown in the preceding diagram:

- **IPsec Traffic**: Securely encrypts the traffic between the clusters using the IPsec protocol.

- **VLAN Traffic**: Enables communication between the clusters using VLANs over layer 2 networks.

- **Cluster Nodes**: The nodes within the Kubernetes clusters that are interconnected by Submariner.

- **Broker**: Handles discovery and connectivity among clusters.

- **Submariner CRD**: A Kubernetes custom resource definition that defines the global Submariner configuration.

- **Submariner Route Agent**: Manages the routing table entries on the cluster nodes.

- **Submariner Gateway Engine**: Establishes secure inter-cluster communication. The benefits of using Submariner include the ability to easily connect services running in different clusters, regardless of their physical location, and the ability to leverage the resources of multiple clusters to scale and balance workloads. This can be useful in a variety of scenarios, such as multi-cloud deployments and disaster recovery.

Limitations and disadvantages

However, Submariner does come with some limitations and disadvantages:

- One such disadvantage is that it requires a high level of network knowledge and configuration in order to set it up and maintain it properly. Submariner requires configuring VPN tunnels between clusters, meaning that it can be challenging to deploy and manage, especially if you are not experienced with VPNs.

- Additionally, Submariner can consume a lot of network resources, which can be an issue if the clusters are running in different regions or even with different cloud providers. This can result in high latency and increased bandwidth usage, which can impact the overall performance of the system. This can be especially problematic in cases where the traffic is sensitive and high-bandwidth, such as in video streaming or large data transfers.

- Furthermore, Submariner may not be well integrated with some existing network management solutions such as firewalls or load balancers, requiring an additional layer of configuration, which can lead to a more complex setup.

There are solutions such as **Red Hat Advanced Cluster Management** (**RHACM**) with Submariner Operator to overcome the shortcomings. You can read about those in the *Further reading* section at the end of this chapter.

Overall, Submariner is a powerful tool for managing cross-cluster communication in Kubernetes, but it does come with certain limitations and challenges.

Skupper – using a common application network (layer 7)

Skupper is software that allows you to connect and expose services running in different Kubernetes clusters or in different environments. It uses a service mesh architecture to provide secure and reliable communication between services, regardless of where they are running. Skupper is built on top of the Kubernetes API, and it can be installed and managed using the Kubernetes command-line tool (kubectl).

As the following diagram depicts, it operates at layer 7:

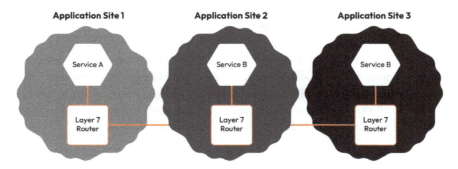

Figure 4.18 – VAN architecture

Skupper handles multi-cluster communication using **Virtual Application Networks** (**VANs**). Let's also try and understand VANs. VANs can provide a seamless connection across multiple cloud environments because they function at the highest level of the networking stack (the application layer). These networks use specialized routers, called layer 7 application routers, to communicate between different applications running on different infrastructure. These routers act as the backbone of a VAN, similar to how traditional network routers are the backbone of a **Virtual Private Network** (**VPN**). However, unlike VPNs, which route data packets between network endpoints, VANs use these routers to direct messages between specific application endpoints, known as layer 7 application addresses.

Skupper architecture

The main components of Skupper are the Skupper control plane and the Skupper edge router. The control plane is responsible for managing service-to-service connections, and the edge router is responsible for managing ingress and egress traffic to and from services. The edge router uses a set of virtual IP addresses and ports that are exposed to the outside world, which are then translated to the appropriate service endpoints inside the clusters or environments.

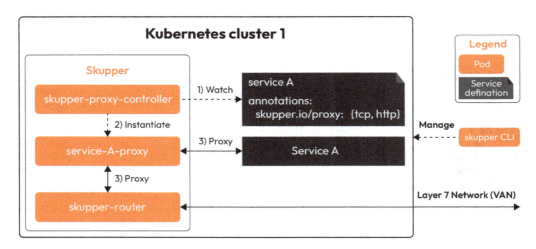

Figure 4.19 – Skupper architecture components

Skupper works by creating a virtual overlay network that connects the different clusters or environments. This is similar to Submariner, but it doesn't require the same level of network knowledge, and it uses a simpler approach. Skupper can be deployed as a Kubernetes add-on, which makes it easy to manage and automate. It also allows you to create a service mesh that spans different environments, which can help with service discovery, load balancing, and secure communication.

I will provide some general guidance on how to set up and use Skupper in Kubernetes:

1. The first step is to install Skupper on your cluster. This can typically be done using a package manager such as Helm or by manually applying the Skupper manifests to the cluster using `kubectl apply`.

2. Once Skupper is installed, you can deploy the Skupper components using the `kubectl apply` command. This includes the `skupper-controller` and `skupper-router` components.

3. Configure Skupper to connect to your other clusters. This typically involves setting up a VPN connection between the clusters and creating a virtual network overlay.

4. Create a service or pod on one of the clusters, then you can access it from other clusters.

5. You can also test the connectivity by running the `skupper status` command to check the connection status and `skupper service-list` to check the list of services that are exposed.

> **Note**
>
> It's important to note that Skupper is an open source tool and there are different ways to install it. Also, it can have changes in the command usage and config file structures in the different versions. Therefore, for a detailed command-by-command explanation, it's recommended to consult the official Skupper documentation (`https://skupper.io/docs/index.html`) for the specific version of Skupper you are using.

Limitations and disadvantages

One of the disadvantages of Skupper is that it only supports Kubernetes, so it cannot be used to connect different orchestration systems together. Additionally, Skupper can be considered less mature and feature-rich than other service mesh solutions such as Istio. Furthermore, the overhead of running Skupper on a cluster could take up resources, which may have an impact on the overall performance of the system.

Service meshes

A service mesh is an adaptable infrastructure layer designed for microservices applications, which enhances the speed, reliability, and flexibility of communication between service instances.

In Kubernetes, a service mesh is implemented as a collection of proxies deployed alongside the application code and managed by the orchestration platform. These proxies (or "sidecars") handle the details of service-to-service communication, such as traffic management, monitoring, and security.

The main components of a service mesh are a control plane and a data plane. The control plane is responsible for managing and configuring the proxies, while the data plane is the actual set of proxies that intercept and handle traffic. The proxies use the configurations provided by the control plane to make routing decisions and apply policies. Notice the service mesh generic architecture in the following figure, which depicts the proxy, control plane, and various important components.

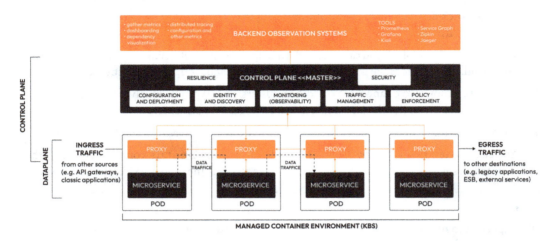

Figure 4.20 – Service mesh generic architecture

When a service mesh is in use, all service-to-service communication goes through the proxies, rather than directly between the service instances. This allows the service mesh to add features such as traffic control, observability, and security without modifying the application code.

The disadvantages of service meshes are that they are complex and can add operational overhead. There is a learning curve for engineers to understand the abstractions that service meshes create and how they work. Also, they increase resource consumption since they will be deployed as sidecars for every service, which means they will consume extra resources. Additionally, they can add latency to service-to-service communication.

Old-school service meshes (classic meshes)

In the open Kubernetes ecosystem, there are several options available for implementing a service mesh, but here are the most commonly used ones:

- Consul
- Istio
- Linkerd

Each of these options has its own advantages and disadvantages, but using any of them can help development and operations teams better manage and maintain microservices. Additionally, many cloud providers and enterprise Kubernetes distributions offer their own managed service mesh solutions that use a combination of these three technologies. For **Telco, Media, and Entertainment (TME)** application stacks, Istio is currently the most popular choice.

Have a look at the Istio service mesh architecture:

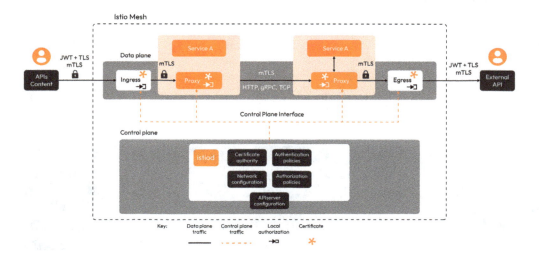

Figure 4.21 – Istio service mesh architecture

In this architecture diagram, you will notice different components and sidecars of Istio at work to enable pod connectivity.

Despite the advantages that sidecars offer over application refactoring, they do not ensure complete separation between applications and the Istio data plane. Consequently, this leads to certain limitations:

- **Tight coupling**: To incorporate sidecars into applications, it's necessary to modify the Kubernetes pod specification and redirect traffic within the pod. As a result, installing or updating sidecars may require restarting the application pod, causing potential disruption for application users.

- **Restricted scope**: Istio's sidecars usually perform traffic capturing and HTTP processing, which are computationally intensive and introduce a latency overhead. Additionally, this process can interfere with applications that use protocols other than HTTP, such as SCTP.

- **Underutilization of resources**: Each sidecar will get provisioned with enough CPU and memory, but over time, it will be underutilized.

A better service mesh (ambient mesh)

The ambient mesh is in its early stages and is designed to separate traditional mesh functionality into two layers. The bottom layer provides a secure overlay for traffic routing and zero trust security. The top layer, which provides access to the full range of Istio features, can be enabled when necessary.

Although the layer 7 processing mode is more resource-intensive than the secure overlay, it still operates as an ambient component of the infrastructure, without any changes required to application pods.

Figure 4.22 – Sliced new mesh

The ambient mesh offers a layered approach for the adoption of Istio, allowing for a smooth transition from no mesh to the secure overlay and full L7 processing on a per-namespace basis, depending on the needs of the user. Workloads in different modes or with sidecars can easily interoperate, and capabilities can be mixed and matched as required, based on evolving needs.

The ambient mesh relies on a shared agent installed on each Kubernetes cluster node, which serves as a zero-trust tunnel (or ztunnel) responsible for securely connecting and authenticating mesh elements. The node's networking stack routes all traffic from the workloads involved through the ztunnel agent, ensuring complete separation of Istio's data plane and application concerns. The ztunnel agent doesn't perform any L7 processing on the workload traffic, which leads to a more efficient and cost-effective approach than using sidecars. This simplified method enables the delivery of shared infrastructure.

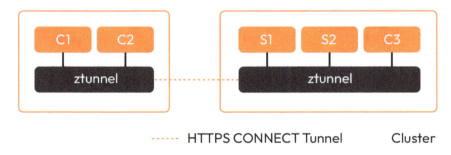

Figure 4.23 – Ambient overlay

Ztunnels form the fundamental framework of a service mesh, providing a zero-trust environment. Once the ambient mesh is activated for a namespace, it generates a secure overlay that enables workloads to use features such as mTLS, authentication, authorization, and telemetry without examining L7 traffic. When L7 capabilities are required, the namespace can be allowed to use them, and waypoint proxies are utilized to handle L7 processing for applications in that namespace.

The service mesh control plane is responsible for setting up the communication pathways, called **ztunnels**, within the cluster to handle all network traffic that requires advanced processing. These pathways, known as waypoint proxies, are essentially regular pods that can be automatically scaled like any other deployment in Kubernetes. This design allows for better resource utilization, as the waypoint proxies can be scaled based on the actual traffic demands of the namespaces they serve, rather than having to anticipate and provision for maximum or worst-case traffic loads. Here is a depiction of a ztunnel:

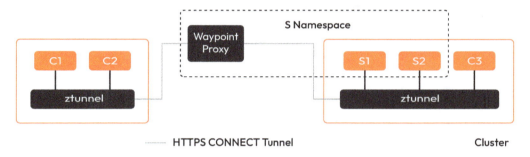

Figure 4.24 – Ambient connect

The ambient mesh uses HTTP CONNECT over mTLS to establish secure tunnels and implements waypoint proxies along the path, utilizing the **HTTP-Based Overlay Network Environment** (**HBONE**) pattern. This approach ensures more streamlined traffic encapsulation compared to TLS, while also enabling seamless interoperability with load-balancer infrastructures.

Combining sidecars and an ambient mesh within a single mesh does not pose any limitations or security concerns for the cluster. The Istio control plane ensures that policies are implemented consistently, regardless of the chosen deployment model. The ambient mesh provides additional options for managing application traffic requirements, with enhanced configurability and a more effective approach.

Federation of service meshes

Service mesh federation is a way to connect and share services and workloads between multiple service meshes, each of which is managed and controlled by its own administrator. By default, communication and information sharing are not allowed between meshes and are only allowed on an explicit opt-in basis. This is consistent with the cloud-era principle of least privilege.

To enable federation, the service mesh control plane must be configured on each mesh to create specific ingress and egress points for the federation and to mark the trust boundary for the overall mesh. Federation also requires additional objects, such as `ServiceMeshPeer`, `ExportedServiceSet`, and `ImportedServiceSet`:

- The `ServiceMeshPeer` object defines the connection between two or more service meshes
- The `ExportedServiceSet` object specifies which services from one mesh can be used by another mesh
- The `ImportedServiceSet` object defines which services are being exported and which are being imported

The following is a depiction of the service mesh control plane:

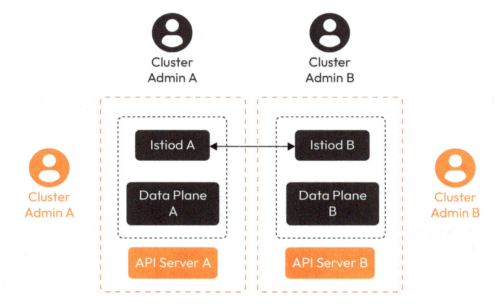

Figure 4.25 – Service mesh control planes to communicate

Service mesh frameworks are dedicated infrastructure layers that streamline the configuration and management of microservice-based applications. These frameworks usually leverage a sidecar container as a proxy, providing features such as mutual TLS-based security, circuit breaking, and canary deployments.

Most popular service mesh solutions also support multi-cluster deployments, allowing traffic to be routed between clusters through a proxy. They can also provide mutual TLS connections across clusters and expose services with features such as cross-cluster traffic routing. However, using such solutions often requires multiple steps and new, specific APIs. The following diagram is a depiction of a service mesh communicating across clusters:

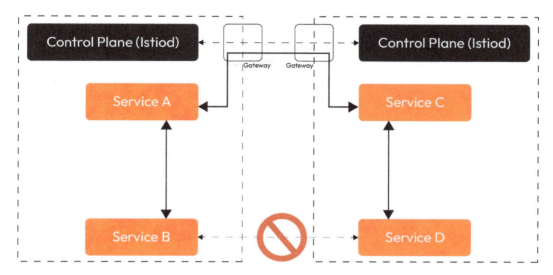

Figure 4.26 – Services communicating via exports and imports

To summarize, service mesh federation enables the management of multiple service meshes across different clusters, allowing for seamless communication and control between services. This approach can simplify the management of complex distributed systems while enabling efficient scaling and resilience. With this, we will summarize this chapter and move on to the next chapter.

Summary

In this chapter of the book, we have explored the communication process between application packages as containers within Kubernetes pods. We have also highlighted the significance of networking as a crucial aspect of any IT solution deployment. Although Kubernetes has introduced new concepts to simplify communication, some of these have both advantages and disadvantages.

To achieve better network performance and DevOps practices, the Kubernetes CNI list has grown substantially. Therefore, it is important to carefully study and select the appropriate CNI type for each application stack. To apply what has been learned in this chapter, it is recommended to evaluate the networking requirements of each application and choose the most suitable CNI plugin for optimal performance and functionality.

In the next chapter, we will learn about design patterns for telecommunications and other industries.

Further reading

- Red Hat Advanced Cluster Management (RHACM):

  `https://access.redhat.com/documentation/en-us/red_hat_advanced_cluster_management_for_kubernetes/2.7/html/add-ons/add-ons-overview#deploying-submariner-console`

- Connect Managed Clusters with Submariner: `https://cloud.redhat.com/blog/connecting-managed-clusters-with-submariner-in-red-hat-advanced-cluster-management-for-kubernetes`

- External-DNS: `https://github.com/kubernetes-sigs/external-dns`

- Links of Multi-NIC Pod use cases: `https://www.redhat.com/sysadmin/kubernetes-pod-network-communications`

Part 2: Design Patterns, DevOps, and GitOps

Part 2 of the book takes a deeper dive into hybrid cloud, exploring the key considerations for operating and securing it. In this section, you will learn about the design patterns that can be applied for operational excellence, ensuring that your hybrid cloud runs smoothly and efficiently. It also focuses on the security of hybrid cloud, examining the various threats and vulnerabilities and the steps that can be taken to mitigate them. Finally, we round off the book by covering the best practices for hybrid cloud, providing valuable insights and recommendations to ensure success. These chapters are a must-read for anyone looking to leverage the power of hybrid cloud and ensure that their implementation is secure, efficient, and effective. Whether you're a seasoned professional or just starting out, you won't want to miss the insights and information shared in these pages.

Here are the chapters that are in this section:

- *Chapter 5, Design Patterns for Telcos and Industrial Sectors*
- *Chapter 6, Securing the Hybrid CloudSecuring the Hybrid Cloud*
- *Chapter 7, Hybrid Cloud Best Practices*

5

Design Patterns for Telcos and Industrial Sectors

In this chapter, we'll explore how **telecom** (**telco**) and other industrial sectors are using a hybrid cloud approach to gain agility and flexibility with their solution deployments. We have distilled the common and reusable elements of various deployments into design patterns. The design patterns can serve as a starter blueprint for a more comprehensive solution that addresses your specific business needs. These design patterns also create a common vocabulary between developers and operations as they design, build, and deploy solutions.

Once you understand the architectural and design principles of these patterns, you will be able to create a design pattern that matches your own unique business needs.

We will cover the following topics in this chapter:

- Applying design patterns for operational excellence
- Creating your own pattern

Applying design patterns for operational excellence

Before we dive into the design patterns for telco and industrial sectors, let's take a look at the underlying reasons for hybrid cloud adoption.

Telco

Telco service providers (also known as communications service providers) are a key part of our everyday lives and provide services consumed across all corners of the world. They include companies, some over a hundred years old, way back from the time of Alexander Graham Bell, with revenue of hundreds of billions of dollars – AT&T, Verizon, Deutsche Telecom, NTT, China Mobile, Telefonica, and Vodafone.

With customers demanding ever more capable communication services, telcos have had increasing budget outlays. By one estimate, the 30 largest telcos incur more than $900 billion in **operating expenditure (OpEx)** and more than $270 billion in **capital expenditure (CapEx)** per year.

Telcos started with **virtual machine (VM)**-based solutions to reduce costs and gain workload placement flexibility by decoupling hardware and software. Containerizing the network workloads further improves operational flexibility and greatly accelerates the speed of deployments.

Cloudification of telcos

Instead of just relying on their own data centers, telcos are increasingly adopting a hybrid cloud approach, enabling them to mix and match on-premises infrastructure with public cloud infrastructure. This approach allows them to deliver across diverse use cases (fixed wireless access, edge deployments, or dedicated infrastructure offerings) with minimal CapEx and OpEx.

This cloudification of telcos is enabled by offerings such as AWS Wavelength, which embeds AWS compute and storage services within telco data centers at the edge of the network. The application traffic can reach application servers in Wavelength without leaving the telco network, thereby reducing latency. In 2021, Microsoft's acquisition allowed AT&T's 5G mobile network to move to Microsoft's Azure for Operators, allowing AT&T to run its core network workloads on the Azure public cloud or in on-premise data centers with common tooling and services.

Telco design patterns

The mobile network can be broadly separated into two domains, with the first one focused on the network itself (its core, access/aggregate layer, and radio access) and the second one focused on support systems. We'll look at three design patterns across these domains:

- **5G core**
- **5G's radio access network (RAN)**
- **Operational support systems/business support systems (OSSs/BSSs)**

5G core

5G is the fifth generation of mobile telephone technology that aims to enable highly immersive experiences and low-latency communication. Every generation of mobile telephony, from 1G to 5G, has continued to advance the capability and capacity of the network with new features. 5G is more than just advancements in wireless communication technology (aka **5G New Radio (5G NR)**) but also the adoption of modular frameworks such as **service-based architecture (SBA)** for its network core. SBA provides control plane functions, delivered as **network functions (NFs)**, that share each other's services over a shared service bus. The NFs communicate with each other using open APIs so NFs from different vendors can be mixed and matched.

5G enables different types of devices connecting per their requested level of performance and characteristics or moving processing closer to the user devices with edge computing. The combination of high bandwidth, low latency, and increased reliability makes 5G well suited to a wide range of applications, example, IoT, autonomous vehicles, or video gaming. A good example of something that telcos need to enable is network slicing. Network slicing is another capability enabled by 5G that allows telcos to create virtual networks, where each slice is customized to customer-specific performance and bandwidth characteristics, over a shared public infrastructure.

5G core, as the name implies, is the central part of a 5G mobile network. It's responsible for establishing end-user connectivity to the network, as well as authentication, policy management, and access to 5G services. The 5G network adopts a **control and user plane separation** (**CUPS**) architecture that separates the control plane and user plane functions of gateway devices. This separation allows the control plane to be located in a centralized location and the user plane to be distributed so that it is closer to the users or applications that it supports. The control plane and user plane functions can be scaled independently of each other, therefore enabling a flexible and efficient core network. Refer to the **3rd Generation Partnership Project** (**3GPP**) (`https://www.3gpp.org/technologies/5g-system-overview`) for details on the scope of 5G core.

Telcos need to offer new services without adding operational complexities and costs. By designing 5G core as a set of disaggregated, **cloud-native network functions** (**CNFs**), telcos can achieve these goals.

Key learning – cloud bursting and driving innovation

Cloud bursting and driving innovation are the key reasons for choosing the hybrid cloud for this use case.

Cloud bursting allows companies to use cloud infrastructure whenever their on-premises resources have reached full capacity. Cloud bursting is suitable for industries that have a fluctuating demand for services, such as retail stores during the holiday shopping season or telcos with high service demand at an event such as a big football game. Instead of over-provisioning on-premises infrastructure to handle unexpected demand for resources or letting the service be degraded due to lack of resources, cloud bursting enables companies to burst the extra workloads to third-party cloud services. The applications that normally run on-premises can be moved to the cloud during peak demand to maintain a seamless user experience. The last few years of COVID provide a good example of cloud bursting, where companies faced a sudden increase in the demand for on-premises infrastructure due to increased use of their digital services or employees working remotely.

5G core solution stack can be categorized into the following:

- Infrastructure

- The application platform

- 5G applications

- Deployment and life cycle management

Here is a conceptual view of a 5G core solution on the hybrid cloud:

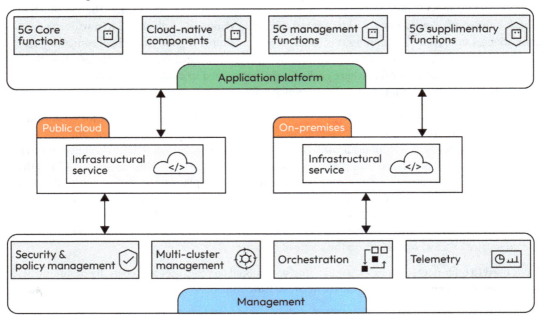

Figure 5.1 – 5G core solution

We will further discuss these components of the stack in the next section.

Infrastructure

This consists of the services and resources provided by cloud service providers and on-premise data centers. These include core resources such as the compute, network, storage, and services such as NTP sources, firewalls, load balancers, and identity management and automation tools. Infrastructure includes bare-metal server instances, VMs, or **infrastructure-as-a-service** (**IaaS**) offerings. The following is an example of the services and resources provided by the infrastructure:

Figure 5.2 – Infrastructure stack example

Each cloud provider has unique service offerings suitable for different workload requirements. For example, AWS Lambda is a popular choice for serverless computing while Azure is popular with companies already using Microsoft tools such as SharePoint or Outlook. These service offerings and infrastructure can be accessed by an **application programming interface (API)** unique to each cloud. For telcos looking to use multiple cloud vendors, these API variations add complexity as they require IT operations to continuously stay abreast of the technology stacks across the data center and various cloud providers. Application platform management tools can help overcome this complexity by adding an abstraction layer that normalizes access across various public cloud providers and data centers.

The application platform

The application platform provides the foundation for 5G applications and NFs to run independently of the underlying infrastructure. Container-based application platforms include container execution, orchestration platforms, and the associated tooling for metrics and dashboards (such as Prometheus). Kubernetes is the most popular platform for deploying applications in the hybrid cloud, keeping them running, and ensuring they can scale based on varying user demand. The following is an example of the services and tooling provided by the Kubernetes-based application platform:

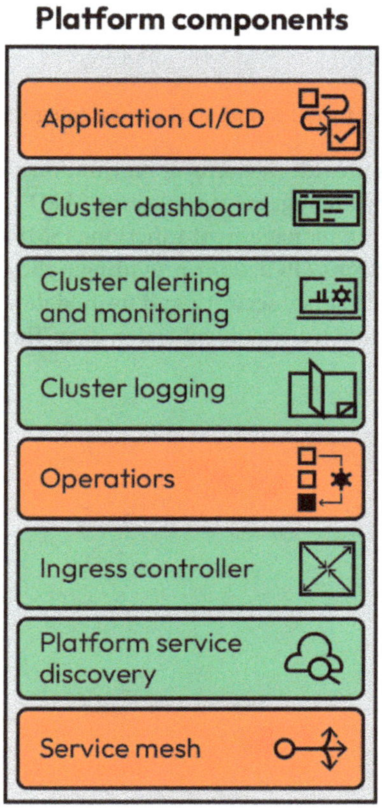

Figure 5.3 – Application platform stack

> **Note**
>
> In 5G core, NFs are microservices that provide capabilities for mobility management, routing, security, policy control, session management, charging, and subscriber data management. Every NF offers its services to all the other authorized NFs through a common API.

The application platform needs to be distributed, **highly available (HA)**, and resilient to allow for the on-demand deployment of 5G applications. In order to allow **continuous integration/continuous delivery (CI/CD)** workflows, the application platform also needs to support features such as automated deployment, intelligent workload placement, dynamic scaling, seamless upgrades, and self-healing.

Application platforms need to support telco-specific requirements that are different from enterprise needs. For example, a Kubernetes Ingress that interacts with other 5G core elements will need to support protocols such as **Stream Control Transmission Protocol (SCTP)**, Diameter, and GTP. In order for Pods to have multiple network interfaces, custom **container network interface (CNI)** plugins such as Multus can be used.

5G applications

5G applications provide the core 5G functionality and can be categorized as follows:

- **Main functions**: These NFs provide device/user registration and session management. These include **access and mobility management functions (AMFs)** for handling connection and mobility management, **session management functions (SMFs)** for interacting with the data plane, **policy control functions (PCFs)** for providing policy rules, **authentication server functions (AUSFs)** used to facilitate security, and **unified data management (UDM)** for the management of user subscription and authentication data. The following is an example of the main NFs:

Main functions

Figure 5.4 – 5G main function stack

> **Note**
> 3GPP standards define the specifications for radio access, the core network, and service capabilities for mobile telephony.

- **Supplementary functions**: These network functions facilitate device/user subscription and value-adding services such as a **unified data repository** (**UDR**), which stores subscriber information that can be used by UDM or the PCF, a **network slicing selection function** (**NSSF**) for the selection of network slice instances, and a **network repository function** (**NRF**) for NF registry and discovery.

The following is an example of supplementary NFs:

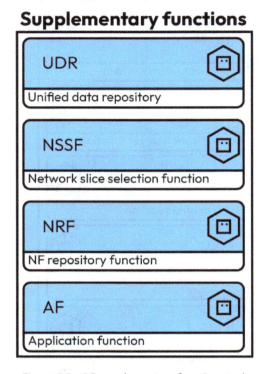

Figure 5.5 – 5G supplementary function stack

The management functions manage different NFs, with an **element management system** (EMS) for managing NFs, a **container network function manager** (CNFM) to manage different types of NFs, and an OSS to manage NFs across the whole network.

All of these 5G applications run on the application platform, spanning multiple clouds and on-premises data centers for resilience and higher availability.

Deployment and life cycle management

As listed in *Chapter 1*, the solution stack deployed on the hybrid cloud needs to be supported with management tools (a hybrid cloud manager), security and policy management tools (RBAC), operators, and a service mesh. The following is an example of the components needed for the deployment of infrastructure, the application platform, 5G applications, and life cycle management (day-2 operations):

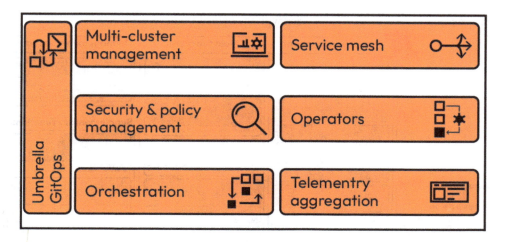

Figure 5.6 – Deployment and life cycle management

These components monitor the hybrid cloud's performance by tracking key metrics such as resource utilization, availability, and response times across various clusters. They're also responsible for multi-cluster management, GitOps, and security.

5G core solution stack summary

Telcos can build a flexible and efficient 5G core by adopting a CUPS architecture that separates the control plane and user plane functions. These functions can be deployed as cloud-native applications or NFs that can communicate over standard interfaces. Each NF can scale dynamically and independently of other NFs, in accordance with the demand for service and the availability of resources.

This allows telcos to support a wide range of 5G use cases without additional operational complexities and costs. By leveraging application platform management tools, telcos can easily orchestrate, operate, and monitor each layer of the network.

Here is an example of a 5G core solution stack with the hybrid cloud (see *Figure 5.7*).

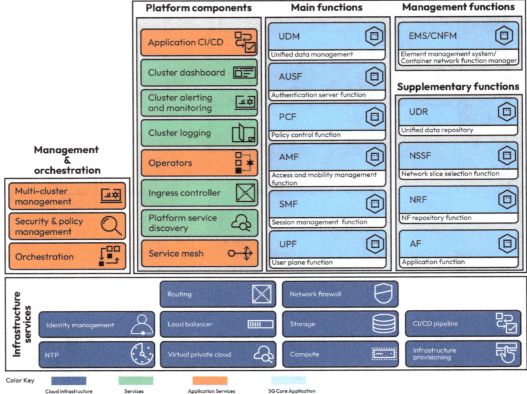

Figure 5.7 – Telco 5G core technology stack

The 5G RAN

A RAN is responsible for connecting mobile devices to a service provider's network. The mobile devices connect to antennas atop radio towers that transmit to base stations where the signal is digitized and routed to the core network. These mobile devices, also known as **User Equipment** (**UE**), can include cars, drones, agricultural machines, or even non-mobile devices such as surveillance cameras or industrial equipment (see *Figure 5.8*).

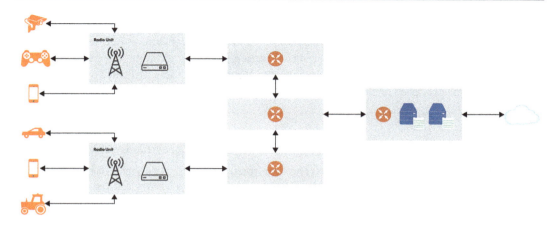

Figure 5.8 – Mobile network overview

As mentioned earlier in this chapter, Telcos incur significant expenses (hundreds of billions of dollars) to keep up with the pace of mobile technology upgrades (see *Figure 5.9*). The RAN accounts for a significant portion of this recurring expense (both in terms of CapEx and OpEx):

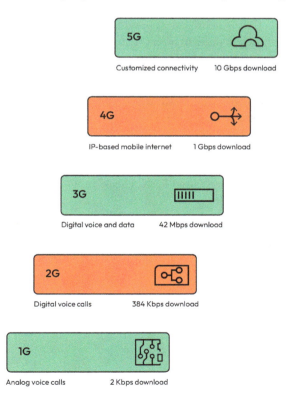

Figure 5.9 – Network technology evolution

Telcos need to simplify network operations to improve efficiency and reduce costs. Traditionally, telcos depended on dedicated network equipment that delivered line speed capabilities but was expensive to procure and could not be quickly repurposed to serve other NFs. In addition, it required narrow but deep technical know-how to configure, operate, and troubleshoot these dedicated systems with ticket-based manual processes, which were time-consuming.

What is 5G RAN?

5G is defined as any system using 5G NR as per the 3GPP industry consortium's 5G standards.

5G can enable dedicated private networks for large customers such as factories or hospitals that meet customers' unique requirements – for example, the 1-ms latency needed for real-time applications. The base stations' capabilities can be customized to match the demand – for example, for less populated rural areas versus densely populated cities.

5G can be implemented as low-band, mid-band, or high-band with frequencies ranging from 60 MHz to 47 GHz. The millimeter waves allow for a smaller cell size, enabling increased connection density per square mile.

For 5G RAN, telcos need to disaggregate the hardware from software with RAN applications deployed as a VNF or CNF (see 5G Core pattern above for a description of a CNF). Separating the control plane from the user plane (a CUPS architecture) provides deployment flexibility, allowing them to scale independently of each other. In order to meet the strict network latency requirements, baseband functions need to be split into **radio units** (**RUs**) for real-time latency, **distributed units** (**DUs**) for near real-time latency, and **central units** (**CUs**) for non-real-time latency.

In order to disaggregate the hardware from software and gain flexibility with workload placement, telcos first started with VM-based solutions (VRANs). With the evolving demands of various 5G use cases, a cloud-native, container-based solution has become a more optimal choice. A cloud-native, container-based RAN solution enables dynamic scaling to changes in demand, lower costs, and ease of upgrades. Additionally, telcos are looking to mix and match RUs from different suppliers by adopting Open RAN. For more information on Open RAN, check out O-RAN ALLIANCE (https://www.o-ran.org/).

Telcos are partnering with cloud providers to offer private wireless networks for specific companies – for example, Deutsche Telekom has partnered with AWS to provide a customized campus network.

Here is the conceptual view of a RAN solution with the hybrid cloud:

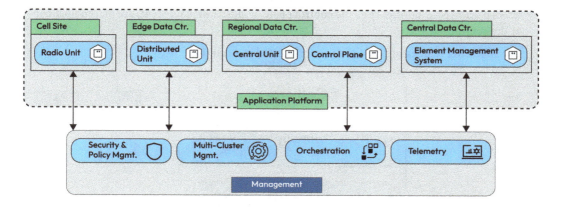

Figure 5.10 – RAN solution

The RAN solution stack can be broadly categorized into the following:

- Distributed infrastructure
- The application platform
- RAN applications
- Centralized management and orchestration

Distributed infrastructure

Unlike 5G core, which is primarily located at centralized sites, the RAN infrastructure is distributed across a number of sites:

- The cell site
- The edge site
- The regional site
- The core site

The choice of site for a workload is determined by the user requirements:

- **The cell site**

    The cell site is the last mile location for telcos' and hosts' DUs, RU, and gateways. The cell site infrastructure needs to allow for distributed deployments of the DU and the RU. The infrastructure for the cell site needs to provide adequate resources to host the workload but is constrained by power, cooling, and space. A distributed architecture approach requires only the resources needed for workload execution to be located at the cell site. This means other needed services or utilities (e.g., the control plane or cluster management) are placed upstream at the centralized site.

The following is an example of the services hosted at a cell site:

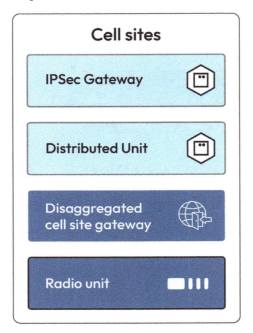

Figure 5.11 – Cell site

- **The edge site**

    Located at the access layer of the network, this site hosts the DU pools, gateway, and time synchronization for the RU and DU deployed at the cell sites.

    The following is an example of the services hosted at an edge site:

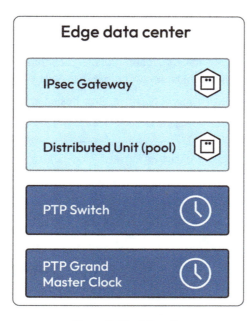

Figure 5.12 – Edge site

- **The regional site**

  Located at the aggregation layers of the network, this site hosts CU pools and gateways and supports functions such as image repositories and identity management to allow the deployment of DUs across edge sites.

  The following is an example of the services hosted at a regional site:

Figure 5.13 – Regional site

- **The core site**

    This site is the core of the mobile network (see the 5G core pattern), hosting the EMS for the RAN components (RUs, DUs, and CUs). It also provides functions such as cluster management, IP address management, and so on to support other distributed sites.

    The following is an example of a subset of the services hosted at the core site:

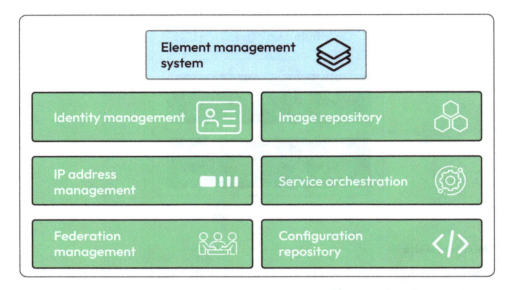

Figure 5.14 – Core site

Decoupling the hardware and software components allows the distribution of RAN components across geographical areas and brings intelligence to the RAN.

The application platform

Similar to the requirements listed in the 5G core pattern, the application platform provides the foundation for NFs to run independently of the underlying infrastructure at various sites. The platform should allow any combination of centralized, distributed, or monolithic deployments.

The application platform needs to be to be configurable for unique needs of each site. This could mean an integrated control plane at a regional site but only remote worker nodes for an edge site (with the control plane located at the core site).

It should be able to support telco-specific features to handle different protocols such as SCTP to guarantee the delivery of signaling messages between the AMF and 5G nodes (N2).

Other characteristics of the application platform remain similar to those in the 5G core pattern – for example, it needs to be distributed, HA, and resilient to allow on-demand deployment of RAN applications.

RAN applications

The various components hosted in the *Distributed infrastructure* section are the RAN applications. These virtualized/containerized applications provide functionality traditionally handled by specialized hardware and software. This includes tools such as xAPPs used by **RAN intelligent controllers** (**RICs**) to manage NFs in real time or rAPPs for automation and management in near real time. *Figure 5.15* shows an example of apps running in a RIC to control the distributed RAN elements, the RUs, DU, and CU.

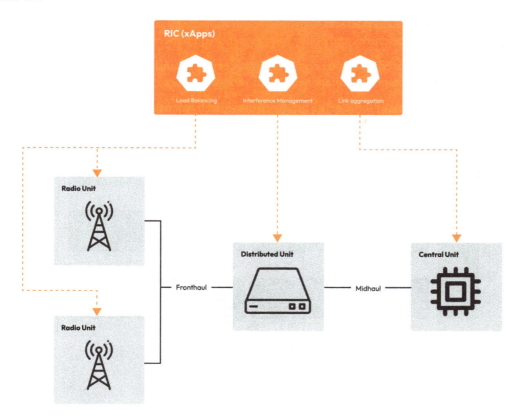

Figure 5.15 – xApps controlling RAN elements

Management and orchestration

In the previous sections, we've seen the many benefits of building a multi-vendor RAN solution with the hybrid cloud. However, this approach brings a whole new level of complexity for telcos. Despite this complexity, telcos are expected to complete new service requests in a few minutes instead of in days or weeks, which was once how long it used to take. By using automation and end-to-end service orchestration, telcos can achieve their service rollout goals. Combined with **artificial intelligence (AI)/machine learning (ML)**-based analytics, this approach can help telcos gain insights to deliver optimized service management and orchestration. The next pattern on OSSs/BSSs shows how telcos optimize network operations with service orchestration.

Telcos also need single-pane-of-glass visibility into the key performance metrics for distributed infrastructure, the application platform, and RAN applications. Any service deployment needs to comply with the necessary security and policy management requirements.

By using open source-based solutions compliant with certain standards, telcos can manage a heterogeneous solution comprising components from various vendors.

The following is an example of some of the functions provided by management and orchestration:

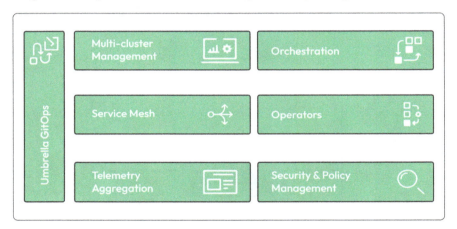

Figure 5.16 – Management and orchestration

5G RAN solution stack summary

By architecting the 5G RAN as a set of disaggregated, cloud-native applications that are distributed geographically, telcos can make it flexible and scalable without adding operational complexities and cost. Each application can scale independently with its placement determined dynamically by the demand for the service and the availability of resources. Telcos use application platform management tools, to orchestrate, operate, and monitor the RAN.

Here is a look at what would a solution stack for a hybrid cloud 5G RAN solution would look like (see *Figure 5.17*).

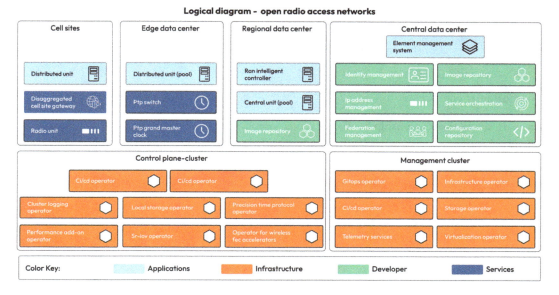

Figure 5.17 – RAN technology stack

Telco OSS/BSS pattern

Telcos use **OSSs** to monitor, operate, and maintain their communication networks and **BSSs** to operate and monetize their services.

The OSS domain covers the following functions:

- Asset and inventory management
- Fault management
- Performance management
- Configuration
- Service provisioning and assurance
- Network planning

BSSs enable telcos to interact with their customers about ordering, billing, and complaints using customer relationship management.

The BSS domain covers the following functions:

- Order services
- Billing
- Customer relationship management
- Customer notifications
- Customer complaints

Figure 5.18 shows the relationship between various components in a 5G solution:

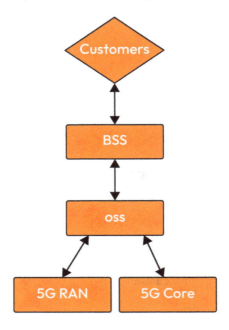

Figure 5.18 – 5G solution with OSS/BSS

Telcos need OSSs/BSSs to gain operational efficiency, reduce costs, add agility, and help create new revenue streams. Traditional OSSs/BSSs were collections of disparate solution architectures and computing platforms. Instead of monolithic OSS/BSS solutions that are deployed in their own data centers, telcos are increasingly adopting a modular OSS/BSS solution distributed across the hybrid cloud, resulting in a faster time to market at a lower cost.

The use of open source frameworks and standard APIs allows telcos to use more **commercial off-the-shelf (COTS)** components sourced from different vendors. TM Forum's (https://www.tmforum.org/oda/) **Open Data Architecture (ODA)** provides a blueprint for a service-oriented, highly automated, and highly efficient approach to OSSs/BSSs. It defines the approach for next-gen OSSs/BSS, comprising an architecture framework, design principles, and specifications for standardized software components based on open APIs.

The OSS/BSS solution stack can be categorized into the following:

- Distributed infrastructure
- The application platform
- OSS/BSS applications
- Autonomous operations

Distributed infrastructure

Similar to the 5G RAN pattern, the infrastructure for OSSs/BSSs also needs to be deployed across a number of sites:

- The core site
- The edge site
- The far edge
- The device edge

The choice of site for an OSS/BSS's workload is determined by the use case requirements:

- **The core site**

  A data center or multi-region public cloud infrastructure is used for deploying OSS/BSS applications that can benefit from highly scalable, HA compute, storage, and network resources. Billing, CRM, and accounting are examples of BSS applications that can benefit from the scalable infrastructure provided by the core site.

- **The edge site**

  OSS/BSS applications interacting with 5G functional components such as an EMS that are deployed at edge locations can benefit from public cloud local zones.

- **The far edge**

  In order to provide low latency, these components are located on-premises. This includes applications, probes, and agents that need to interact with network components. xApps and rApps are examples of network automation tools used to increase the operational efficiency of the RAN. xApps tools are used by the RIC in real time to optimize radio spectrum utilization while rApps are used to manage non-real-time events within the RIC.

- **The device edge**

  This includes interaction with UE-like IoT devices monitoring the flow of data from these devices to the OSS/BSS core.

The application platform

Similar to the requirements listed for the 5G core and 5G RAN patterns, the application platform provides the foundation for OSS/BSS functions to run independently of the underlying infrastructure at various sites. The platform should allow for any combination of centralized, distributed, or monolithic deployments.

The application platform needs to offer different deployment options to be configured to meet the unique needs of each site. This is similar to application platform requirements for 5G RAN pattern we saw earlier.

To enable geo-redundancy and fault isolation, there should be direct networking between Services and Pods in different Kubernetes clusters across the hybrid cloud. An open source project such as Submariner provides this kind of cross-cluster L3 connectivity.

Other characteristics of the application platform remain similar to the 5G core and 5G RAN patterns in that it needs to be distributed, HA, and resilient to allow on-demand deployment of OSS/BSS applications.

OSS/BSS applications

The various components hosted in the distributed architecture section are the OSS/BSS applications. These virtualized/containerized applications provide functionality traditionally handled by specialized hardware and software. This includes tools such as xAPPs and rAPPs.

By using a modular approach, generic functionality such as fault management or performance management can be combined in a unique manner to allow telcos to offer differentiated services. However, from an operational perspective, such a modular approach brings its own complexities. To overcome this challenge, each OSS/BSS module (microservice) can be implemented as an abstraction layer over an **EMS** – refer to the 5G core pattern. The EMS abstraction layer enables real-time observability with logs, metrics, and traces across various 5G network functions (CNFs). A 5G CNF, such as the AMF, SMF, or PCF can provide a **charging trigger function** (CTF). The CTF provides the necessary information to a **5G charging function** (CHF), a part of the **converged charging system** (CCS).

The CHF enables telcos to charge users for the services consumed by collecting network and service usage data. The CCS brings together both the online and offline charging systems so telcos can appropriately monetize various 5G use cases. The monetization should correlate to the value offered through the quality of service, network slice, or IoT data stream. CCS is integrated with the BSS operations flow via an API gateway. Network segmentation and isolation are important for OSSs/BSSs as these business-critical services should not be directly connected to the internet. With the hybrid cloud, we need to use private networks and API gateways or **network address translation (NAT)** to access external cloud services. The charging system should simplify the process of creating new offers and services through automated workflows.

Open source technology is key to building an infrastructure-agnostic OSS/BSS solution. Tools such as Istio service meshes, Kiali, Jaeger, Zipkin, Prometheus, and Thanos are key ingredients for the backend integration and performance metrics of an OSS/BSS solution.

Autonomous operations

Every component of the solution, whether the infrastructure, platform, or OSS/BSS applications, needs to be independently scaled up/down and managed autonomously.

It's a fairly complex undertaking to operationally manage applications, platforms, and various other components of the solution across the hybrid cloud. Although some components of the solution provide autonomous operations, such as Kubernetes scheduling and life cycle management containerized applications, there are several other aspects of the solution that still need to be automated – for example, configuration and security controls across different cloud environments.

It'll require a combination of technology and processes to meet the automation needs of this solution. Here are a few considerations:

- A cluster management solution to automate deployment and life cycle management of the application platform
- GitOps (e.g., ArgoCD) to declaratively configure infrastructure and applications across different environments
- An automation platform (Ansible or Puppet) to automate the configuration of routers, load balancers, and firewalls
- AI/ML analytics to optimize network operations based on real-time data

OSS/BSS solution stack summary

By architecting a modular OSS/BSS solution that's distributed across the hybrid cloud instead of a monolithic solution deployed in private data centers, telcos can achieve a faster time to market at a lower cost. Similar to the benefits listed for other adoption patterns, every OSS/BSS application can autonomously scale independently of each other, and its placement is determined dynamically by the demand for the service and availability of resources.

Here is a look at what would a technology stack for a hybrid cloud OSS/BSS solution would look like:

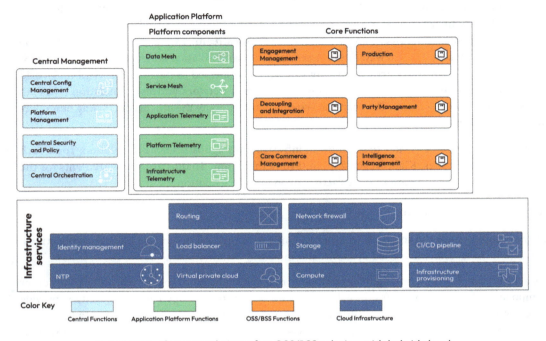

Figure 5.19 – Conceptual view of an OSS/BSS solution with hybrid cloud

An industrial edge pattern

Companies in manufacturing, oil and gas, power plants, and utilities have a dual model for computing infrastructure. While their core business functions are managed by traditional **information technology (IT)**, the operational resources (factories, oil drilling rigs, and power substations) are managed by **operations technology (OT)**. These operational resources have been purposefully segregated to avoid any interference from enterprise IT, but as a result, these systems have typically lagged behind the advancements in IT systems by a decade or more. This lag has deprived the operations side of the agility, cost savings, and flexibility brought to enterprise applications by modern IT approaches. One of the biggest reasons cited for keeping this separation between IT and OT was the belief that OT systems were more secure because they were segregated from IT systems. As much-publicized hacking incidents in the last few years have shown (such as with JBS and the Colonial Pipeline), OT systems are not more immune but have a higher risk profile due to the disproportional impact that shutting down these systems can have on the entire supply chain – for example, gas outages across the east coast of the US and meat shortages resulting in price spikes.

The industrial sector can gain many advantages by adopting modern IT approaches such as hybrid cloud. The best practices of open hybrid cloud that leverage technologies such as virtualization, containers, and automation can also be extended to the industrial edge, all the way to the factory floor or other industrial sites. OT resources in factories can be deployed and managed in a secure and standardized manner.

> **Note**
>
> **Using containers for IoT**: Industrial IoT applications require modularity, isolation, scalability, and resource efficiency, making the use of containers an especially good fit. IoT applications need to be deployed on many different edge tiers, each with its own unique resource characteristics. Combined with DevOps practices and a rich ecosystem of tooling, containers are becoming a popular choice for developing, deploying, and managing IoT applications.

These approaches can enable companies to proactively identify equipment issues, automate workflows to reduce downtime, minimize human errors, and optimize operations. Combined with AI/ML, management can understand and predict the state of its production systems. As evidenced by supply chain shortages during COVID, it's critical for not only industrial companies but also their supply chains to become operationally agile.

Here is the conceptual view of an industrial edge solution with the hybrid cloud:

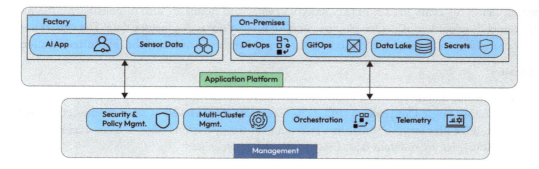

Figure 5.20 – Industrial edge solution

For the industrial edge pattern, we'll look at the example of a large manufacturer with dozens of factories in geographically dispersed locations, that is, edge sites. The on-site systems at the factory receive sensor data from operational systems (e.g., a robotic paint arm) and analyze it in real time with AI/ML models to predict possible equipment failures. The aggregated sensor data for all the factory systems also need to be sent to the data center or the cloud for further processing and storage. In addition, the hardware and software systems at the factory also need to be deployed and managed throughout their life cycle.

This entire process workflow needs to be consistently repeated at dozens of factory sites without tedious and error-prone manual configuration. For utilities or oil and gas companies, this workflow needs to extend to hundreds or even thousands of on-shore or off-shore drilling sites. The industrial solution stack can be categorized into the following:

- Distributed infrastructure
- The application platform
- Deployment and life cycle management

Distributed infrastructure

Similar to the 5G RAN and OSS/BSS patterns, industrial infrastructure is distributed across the following:

- Centralized sites
- Edge sites

The choice of site for a workload is determined by the characteristics of the workload and the available computing resources.

Centralized sites

In addition to handling the core enterprise functions (e.g., an ERP system, payroll, and asset management), the centralized sites (e.g., the data center and cloud) are also responsible for the AI/ML model development, DevOps pipeline, and long-term storage of factory data. The cluster and application life cycle management functions, for all sites, are also handled at the centralized location.

Edge sites

Edge sites include both the physical infrastructure, such as assembly line equipment, boilers, turbines, **Programmable Logic Controllers** (**PLCs**), sensors/actuators as well as servers and networking equipment. Edge sites are also sometimes referred to as the operations edge. The considerations for the operations edge are different from those for the other use cases:

- **Data deluge**: IoT devices generate a continuous stream of data at massive scales – for example, refineries generate 1 TB data per day, while aircraft engines generate 1-2 TB data per flight. It would be prohibitively expensive to send all this data (most of which is redundant and not actionable) to a centralized site for further processing. A preferred approach is for telemetry data from sensors to be normalized, processed, and aggregated at these edge sites before being sent to a centralized location.

- **Latency sensitivity**: For functions where critical decisions need to be made in real time (e.g., the control function of robotic arm), the data processing needs to happen at these edge sites in real time to avoid latencies in communicating with the centralized site.

- **Business continuity**: The core business functions of a factory, hospital, or retail store need to continue to operate even when the connectivity to a centralized site is degraded or lost. The factory should also host source code repositories so that apps at edge sites can continue to be provisioned while disconnected from the centralized site.

- **Distributed workloads**: As new capabilities become available at edge sites, the workloads can be distributed across edge and centralized sites. As an example, AI/ML models for predictive maintenance of factory equipment are developed at the centralized site, while real-time inferencing can be done at the edge site.

The technology stack for device edge needs to account for unique conditions such as the following:

- **Industrial design hardware**: Low-power industrial IoT gateways designed to withstand harsh environments

- **Software**: A small-footprint operating system that can be updated remotely

- **Non-manual processes**: Autonomous application deployment and life cycle management

Cloud service providers also have offerings such as AWS Outposts or Azure IoT Edge, which provide pre-built compute and storage hardware that allows customers to deploy compute and storage at the operations edge while still being able to use cloud services.

The application platform

Similar to the requirements for the other patterns, the application platform provides the foundation for industrial edge applications to run independently of the underlying infrastructure at various edge sites. The computing infrastructure at edge sites varies greatly, with applications running on decades-old legacy systems, virtualized solutions, or containerized environments. The application platform should allow for any combination of centralized, distributed, or monolithic deployments of applications.

The application platform characteristics need to be configured to meet the unique needs of each site. This means HA application platform clusters are deployed at more capable sites (e.g., in a large factory) but scaled down to a single or dual system when deployed at less capable sites (e.g., a windmill or retail kiosk). Similar to the technology stack requirements for an edge site, the application platform needs to match the unique needs of the device edge – for example, Red Hat MicroShift or Rancher K3s field-deployed edge computing devices.

Other characteristics of the application platform remain similar to the former patterns in that it needs to be distributed, HA, and resilient to allow for the on-demand deployment of industrial applications.

Deployment and life cycle management

Industrial sites lack the IT resources and skills for troubleshooting any issues with their computing infrastructure; hence, edge sites need to be entirely managed from a central location. This includes not only initial deployment but also subsequent updates/upgrades of applications and the underlying infrastructure.

Similar to the OSS/BSS pattern, each component of the solution, whether the infrastructure, the platform, or industrial applications, needs to be independently scaled up/down and managed autonomously.

Key learning – edge computing, business continuity, data protection, and data privacy

Edge computing, business continuity, data protection, and data privacy are the key reasons for choosing the hybrid cloud for this use case.

Edge computing allows companies to bring computing closer to their customers or sources of data. Edge computing enables use cases requiring low latency, such as a robotic arm in a factory or disconnected computing, where remote sites need to continue to operate without communication with a central site. For telcos, edge computing can help improve the customer experience by moving applications or content toward the edge of the network. With **multi-access edge computing** (MEC), telcos can support a new class of applications and services at the edge of the network to take advantage of their proximity to the customers. MEC can enable low-latency gaming or **virtual reality** (VR) applications. However, there is a great variety of edge use cases, each with its own unique requirements – so, an IoT edge for autonomous vehicles is very different from that related to a factory, smart city, or telco.

A hybrid computing model uses a combination of edge computing and centralized computing, where centralized computing is used for compute-intensive workloads, while edge computing is used to solve problems at the source in near real time.

Solution architects need to evaluate the requirements for each use case taking into account the expected benefits of edge computing. If the use case doesn't require reduced latency, disconnected computing, or other attributes, then edge computing may not be the right choice.

No modern business can continue to function without its digital infrastructure consisting of servers, applications, databases, and so on. Any impact on this digital infrastructure, whether due to natural disasters or ransomware attacks, would cause disruption to business operations as well as severe financial and reputational damage. The public cloud could be used as an offsite backup option for on-premises infrastructure or backing data up in different regions across multiple public clouds could add resilience. Instead of backing up everything (which becomes cost-prohibitive), companies should identify the critical data and applications that should be backed up frequently.

Sensitive data consists of important or confidential information that needs to be protected from unauthorized access such as data with high operational or business value (e.g., factory production data or force deployment data for defense forces). This data needs to be processed and stored with higher protections to safeguard it during all stages of its life cycle through generation, processing, transmission, and storage. The government regulations such as the **General Data Protection Regulation** (GDPR) set specific requirements for processing and storing sensitive data.

Industrial solution stack summary

By adopting modern IT approaches such as hybrid cloud, companies in industrial sectors (Siemens, Caterpillar, Exxon Mobil, etc.) can manage their OT resources in a secure and standardized manner.

This enables these companies to proactively identify equipment issues, automate workflows to reduce downtime, minimize human errors, and optimize operations. Combined with AI/ML, management can gain new insights into the state of their production systems.

Here is a look at what a technology stack for a hybrid cloud industrial solution would look like (*Figure 5.21*).

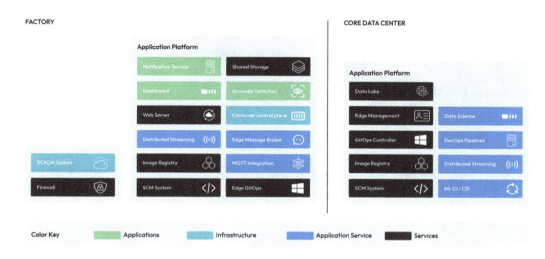

Figure 5.21 – Industrial solution stack

Creating your own pattern

Now that we've seen how telco and industrial segments use hybrid cloud for deploying solutions for their use cases, let's look at how you can use the hybrid cloud to deploy a solution applicable to your own use case. You probably noticed some common themes and elements across the telco and industrial design patterns. These commonalities provide you with a blueprint to create a design pattern for a use case of your own choosing.

Here are some of the things to consider when creating your design pattern.

Defining a framework

In order to make the portability of applications across multiple computing environments possible, we need to have a unified platform for orchestration and management across private and public clouds. In the past few decades, virtualization provided hardware abstraction so the applications could run across disparate environments. However, the need for agility is driving the adoption of microservices and cloud-native technologies such as containers, Kubernetes, CI/CD), and GitOps.

As the adoption of microservices-based architecture accelerates service delivery, this often translates into increased market share. Check out the 6R framework in *Chapter 1* for more details.

Here is a conceptual view of a microservices-based architecture deployed on a hybrid cloud:

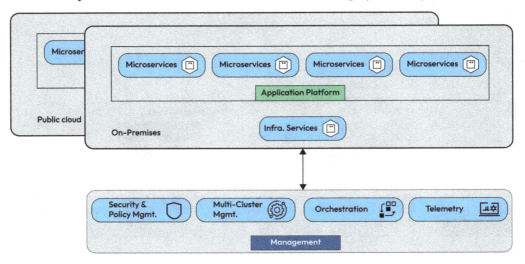

Figure 5.22 – Microservices-based solution

Every hybrid cloud environment is different as defined by the unique needs of the customer. There are some common elements that are consistent across multiple use cases. Some of these elements are as follows:

- Cloud-friendly
- Uses a common application platform
- Involves consistent management
- Uses automation

Cloud-friendly

Unless your company is less than 10 years old, there is a good chance for an existing business to have existing infrastructure and applications that predate cloud technology. These legacy applications and infrastructure are likely responsible for running core parts of the business (e.g. the ERP system, batch storage, or operations) using highly optimized and interconnected systems that may not be suitable for a quick migration to the cloud. As an architect, you need to take inventory of what business processes can benefit from cloud-native technologies. This technical analysis happens only after the business analysis comes out in favor of using the cloud. The ideal applications for the cloud are based on the 12-factor app methodology (https://www.redhat.com/architect/12-factor-app).

If the business needs align with migrating to the cloud, consider refactoring the app so its various components can be distributed and scaled independently (as opposed to containerizing the legacy monolithic app and hosting it in the cloud!).

Not only are emerging applications such as IoT, AR/VR, robotics, and telco network functions well suited to the hybrid cloud but also even traditional enterprises can benefit from the hybrid cloud model to better support geographically dispersed locations such as remote/branch offices or retail stores.

A common application platform

For applications to be seamlessly deployed across the public cloud and on-premises, a consistent application platform such as Kubernetes is needed. By using the same software stack, it's easy to extend the environment from the public cloud to on-premises or vice versa. Platform offerings from AWS Outposts, Azure Arc, Google Anthos, VMware Cloud, and Red Hat OpenShift all enable seamless portability across on-premises environments and the cloud.

Here is a conceptual view of a common application platform across the hybrid cloud:

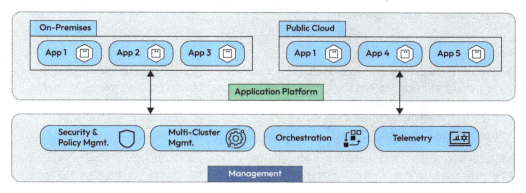

Figure 5.23 – Common application platform

Consistent management

Both the cloud and on-premises infrastructures consist of many moving parts, which need to function together seamlessly all the time to deliver a consistent service experience to users. There needs to be a unified approach to provide full visibility and control of infrastructure, applications, services, and users across both on-premises and cloud environments. The telemetry data about the state of each component, performance monitoring, security auditing, reporting, and analytics need to be collected and provide a single-pane-of-glass view across infrastructure, applications, services, users, and costs. In addition to cloud provider management tools such as the AWS Management Console and the Azure portal, there are also third-party tools such as Nutanix Prism Element, CloudBolt, Datadog, Red Hat ACM, and so on available.

Automation

Automation is a key consideration for modern IT teams and is used across all aspects of technology stacks and processes. Automation enables efficiency, speed, and accuracy for repetitive tasks. Eliminating repetitive manual tasks reduces the risks of human error and increases productivity, consistency, and reliability.

One of the main reasons why Kubernetes is the preferred application platform for hybrid cloud is because it allows you to automate the deployment, scaling, and management of your applications and infrastructure.

Telcos and their suppliers may want to look at Nephio, an open source project, which provides Kubernetes-based automation of network functions and the underlying infrastructure supporting these functions.

For industries requiring regulatory compliance (e.g., the HIPPA for healthcare or the PCI DSS for the financial sector), automation tools should support the needed audits and security checks. For edge sites that are air-gapped, the tools should have offline capabilities. Red Hat Ansible and Chef are often the tools of choice for automation tasks.

> **Note**
>
> Despite all the benefits of the hybrid cloud, it would be intellectual dishonesty to claim that it's the perfect solution for every situation. If your business is being served well by monolith applications that do not need to scale up or down dynamically or your services do not benefit from regional availability zones, then you may not be ready to move to the hybrid cloud.

As architects, you will need to identify use cases that are aligned with the hybrid cloud. If a use case doesn't benefit from the many advantages brought by the hybrid cloud, then you may need to consider other options.

> **Note**
>
> Although cost is a business imperative, as a technical expert, you'll be required to provide the key analysis for the cost of deploying a solution to the hybrid cloud. In addition to the cost of infrastructure, software, and services, you also need to consider the ingress/egress costs for data and applications located across different infrastructures.

Here is the process you need to follow to create your own design pattern:

1. Identify the business requirements and outcomes desired by your stakeholders. This needs to be translated into technical requirements, user experience, and product features.

2. Identify where each workload and application needs to be deployed to best meet the requirements and goals defined in *step 1*. This could mean frontend applications deployed in the public cloud while backend systems dealing with sensitive business information are hosted in a private cloud.

3. Choose the hybrid cloud infrastructure, application platform, and tools for orchestration, management, and automation that can be deployed on your choice of hybrid cloud defined in *step 2*.

4. Create a migration plan to move your existing applications to a hybrid cloud environment, keeping in mind the performance, cost, and scalability requirements.

5. Create configurations, security compliance, and automation processes to deploy infrastructure and applications to the hybrid cloud.

6. Establish protocols for data management, including storage, access, and backup.

7. Test and validate the application deployments to make sure applications and services function as expected.

8. Continuously monitor the key performance indicators to ensure the applications and infrastructure continue to meet the business and technical requirements.

Summary

This chapter put into practice the concepts covered earlier in the book by looking at design patterns for telco and industrial segments.

Telcos can build a flexible and efficient 5G core by separating the control plane and user plane functions, which can be deployed as CNFs on the hybrid cloud. The 5G RAN can be architected as a set of disaggregated, cloud-native applications that are distributed geographically. Telcos can also build a modular OSS/BSS solution that's distributed across the hybrid cloud instead of a monolithic solution deployed in a private data center. This approach allows telcos to support a wide range of 5G use cases without additional operational complexities and costs. By leveraging application platform management tools, telcos can easily orchestrate, operate, and monitor each layer of network.

The industrial design pattern showed how companies in the industrial sector can manage their OT resources in a secure and standardized manner using the hybrid cloud. This design approach enables companies to proactively identify equipment issues, automate workflows to reduce downtime, minimize human errors, and optimize operations.

These design patterns provide you with a blueprint for creating a design pattern for a use case of your own choosing. Careful attention must be paid when choosing the hybrid cloud infrastructure, application platform, and tools for orchestration, management, and automation. The business needs should guide your technical requirements – for example, where each workload and application should be deployed or whether the needed protocols are supported.

Further reading

- O-RAN specification: https://www.o-ran.org/specifications

- 5G system overview: https://www.3gpp.org/technologies/5g-system-overview

- TM Forum ODA: https://www.tmforum.org/oda/

- Architectural pattern examples, templates, and tools: https://gitlab.com/osspa/portfolio-architecture-examples https://www.redhat.com/architect/portfolio/#

6

Securing the Hybrid Cloud

In the previous chapter, we learned about how customers interested in a hybrid cloud architecture deploy solutions on a hybrid cloud and explored the design patterns in the context of operational excellence. In this chapter, we will go to the heart of all the design principles we have learned thus far and explore the application of security best practices. The fulcrum of any architecture design pattern is security. Particularly with hybrid cloud, security becomes increasingly more complex as it represents the culmination of security principles and practices from an array of different infrastructures, platforms, and application technologies.

It is especially important to understand the nuances of security principles in the world of hybrid cloud as consistency becomes a key consideration. A robust architecture with flexibility helps address all the enterprise requirements and minimize any security risk, thereby efficiently protecting the data, applications, and infrastructure of the hybrid cloud.

In this chapter, we will cover the following topics to understand various principles of security across the key components of hybrid cloud:

- Understanding the core security principles
- Different components of security in the hybrid cloud
- Configuring security within a hybrid cloud environment
- Configuring network security
- Protection of data
- Securing hybrid cloud operations
- Compliance and governance

Understanding the core security principles

To begin, let's examine the fundamental principles of security so we can then see how to apply those principles to the various components of hybrid cloud. It is important for hybrid-cloud architects to take a very holistic approach to security as it has a notable impact on all components of the architecture. The security posture of an enterprise has widespread implications for the business as it touches on digital sovereignty, data protection, compliance, and regulations depending on the type of industry and the business's geographical location.

This brings us to the core security principle, **trust**. Security should not be bolted on as an afterthought, but rather built from the ground up. One very common approach to a built-in model of security is the **zero-trust** approach.

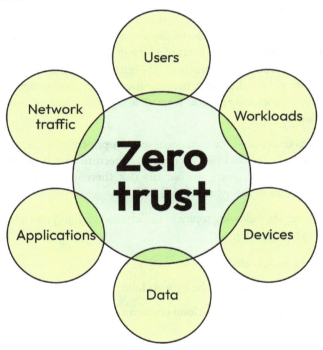

Figure 6.1 – Zero-trust approach to security

In this approach, all the entities of the architecture (and all of the components within those entities) are treated as untrusted by default and thus considered a potential threat. Therefore, each of these resources has to demonstrate its trustworthiness through some form of verification and recursive validation before it can become part of the architecture.

Translating the preceding diagram into hybrid cloud components produces the following illustration, showing the layers of trust to be traversed to protect your systems by eliminating holes, gaps, and flaws, whether in infrastructure or software:

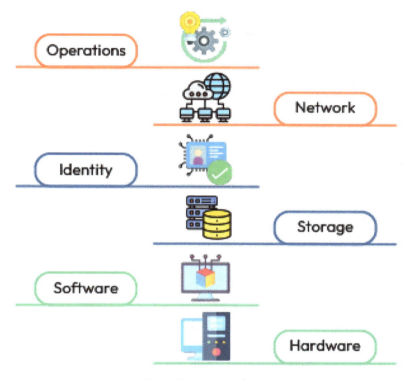

Figure 6.2 – Layers of trust

As exciting as this sounds, zero trust at the hardware layer might not be possible for on-prem infrastructure running legacy workloads. However, we should continue to apply the principle of zero trust to the other layers wherever applicable.

So, let's now take this core principle of security – trust – and start applying it throughout our architecture. In the next section, we will break down various components in our architecture and identify the key aspects that affect security.

Different components of security in the hybrid cloud

The following diagram shows the various components of security across hybrid cloud infrastructure:

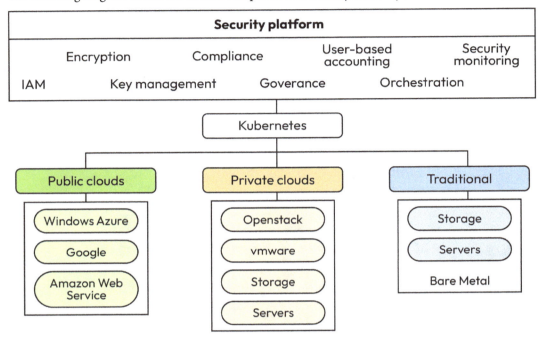

Figure 6.3 – Components of security in hybrid cloud infrastructure

Security must be applied across the entire stack of the application, from the **infrastructure** and **platform** to the end-user **application** itself. Applying this to the hybrid cloud architecture, this stack could be spread across a public cloud, a private cloud, and on-prem infrastructure, and could even be extended further to edge networks and edge devices. The term *edge network* refers to a local area network that connects to the internet and interfaces with public/private clouds or on-prem systems. With the advent of smart devices as part of the IoT, along with more powerful mobile devices, the scope of systems of record and systems of engagement has been significantly expanded. These are some examples of edge devices that could be part of the edge network. *Edge*, as the name suggests, also denotes an entry point into the main systems of our network.

Edge devices expanded the scope of the hybrid environment. Let's now explain the components of hybrid cloud using an example architecture that spans across many entities:

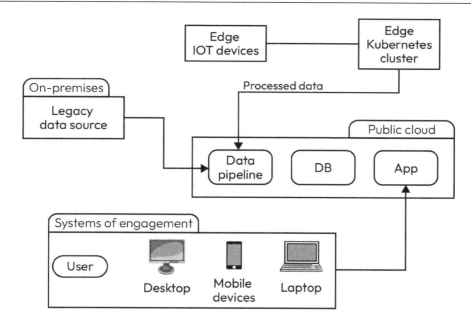

Figure 6.4 – Single hybrid cloud architecture

The preceding diagram shows a simple web application running inside a Kubernetes cluster that can be accessed by users from desktops, laptops, and mobile devices. The application stores its data in a cloud-native database that gets populated through a data pipeline. It receives data from an edge cluster running Kubernetes, which handles the pre-processing of data from IoT devices, and also from a legacy data source running in an on-prem data center that sends data in batches. We will use this example throughout this section to explain how security relates to each of these components.

Within all these entities, we can identify the following components that have some kind of effect on security:

- Hardware
- Boot
- The hardened kernel
- Storage and encryption
- The low-level physical network
- Audit logging
- The guest OS
- Network security

- Platform security

- Platform operations

- Platform **Identity And Access Management (IAM)**

- Deployment and usage

- Application configuration

- AuthN and AuthZ

- Content and data

The main reason for listing these components is that the boundary of responsibility changes from the customer to the cloud provider depending on the application's components running on the respective entities as listed here. For example, the entire responsibility of security shifts to the customer if the component of the application is on-prem. On the other hand, the customer's security overhead considerably reduces if the component is run on a PaaS or SaaS on a public cloud.

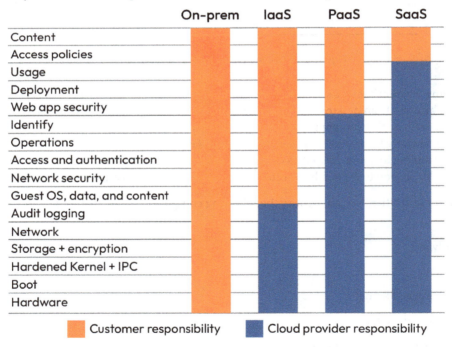

Figure 6.5 – Boundaries of responsibility

The preceding illustration shows how the boundary of responsibility shifts from the cloud provider to the customer as the solution changes from **Software as a Service (SaaS)** to **Platform as a Service (PaaS)**, **Infrastructure as a Service (IaaS)**, and finally to a completely on-prem hosted solution. Kubernetes notably falls into the gray area between IaaS and PaaS and is usually referred to as **Container as a Service (CaaS)**, and is not shown in this diagram. When it comes to Kubernetes, there are a variety of distributions available ranging from open source to vendor-based, and each provider offers different products with different levels of service and functionality, each with their own impact on security.

In the diagram, we can see the infrastructure, platform, and application components. The hardware, boot, hardened kernel + IPC, physical storage, network, and audit logging represent the infrastructure layer. The guest OS, data and content, platform networking, access, auth, operations, and identity make up the platform components. Web app security, deployment, usage, access policies, and content belong to the application layer.

Now that we have understood the entities and components, let's now understand security as we traverse through these layers. It is important to first identify which components of your application belong to which entities in your hybrid cloud architecture. We will continue with the example in *Figure 6.4*.

The infrastructure layer

Let's analyze the two components that are not running in the public cloud. The first component is a legacy data source running in the customer's on-prem data center. Second is the Kubernetes cluster running at the edge location and pre-processing data from IoT devices. Both these environments can be considered on-prem and hence, the customer has the sole responsibility of managing the infrastructure, platform, and application. As mentioned earlier, it's not possible to enforce zero trust at the hardware level for these application components. The infrastructure layer also includes the operating systems installed on the respective servers. The same is true with the Kubernetes layer as well: the operating systems of the cluster nodes will also need to be patched.

One best practice for keeping the Kubernetes nodes' OS secure is to use an immutable operating system for the nodes. An immutable operating system is a lightweight operating system that provides only the required functionality to run your applications and does not include a package manager for updates. As the name suggests, the core elements of the OS are read-only and usually have to be fully replaced when patching or updating the OS components. This significantly reduces the surface area for attacks and helps immensely to increase the security of the operating system. Since these OS distributions are lightweight, it's very easy to patch and upgrade these systems.

Most enterprises, as a best practice, choose an OS distribution from a vendor. It's part of the vendor's SLA to constantly keep the operating system secure. Depending upon the OS distribution, it may be the responsibility of the IT team to make sure the OS is kept properly patched and up to date as regards any vulnerabilities that emerge.

In addition to this, the customer must also validate the security of the hardware used in their IoT devices via their vendor's validation process. This helps avoid things such as Meltdown and Spectre, two of the most recent hardware vulnerabilities discovered:

Figure 6.6 – Example of hardware vulnerability

In the case of public clouds, each cloud provider has their own way of implementing and enforcing zero trust at the hardware level. Some cloud providers use custom-built hardware, or might fully manufacture their own hardware without the involvement of any third parties. Custom-built chips embedded in servers can be used to check the machine's integrity every time it boots. The machine is neither able to join the network nor process any data until the integrity is properly established. Cloud providers could also use computing devices such as hardware security modules or trusted platform modules, which employ cryptographic operations to verify the authenticity of the hardware.

In addition to the core infrastructure components in the cloud, the other cloud, on-prem, and edge components all have their own network security features and capabilities, which are either native to the cloud platform or enabled using third-party vendor solutions as chosen by the enterprise. Enterprises could choose a cloud-agnostic solution to maintain consistency across the entire hybrid cloud platform. We will dive deeper into this area in the *Configuring network security* section of this chapter.

The platform layer

Making our way up into the platform layer, our example contains two platform varieties for which security needs to be considered. First, we have the legacy database platform, which could be anything from Oracle or SQL Server to DB2, MySQL, and so on. Whoever the database vendor, they are responsible for the database platform's security. It's thus important to have applied the latest security patch to the version of the database being used.

The other platform in our example is Kubernetes. To secure this platform, enterprises have to begin by either choosing a community project or a distribution of Kubernetes supported by vendors. This is a very important and impactful choice in terms of the security implications on the Kubernetes layer. Running the community version of Kubernetes in the production environment is quite a heavy lift as the day-two operations continue to complexify as the enterprise scales. This book will not get into which option is better as there are plenty of resources available that advocate for a certain vendor-supported version.

There are quite a few recommended best practices for securing the Kubernetes platform and the applications that run on top of it. We will look at them in detail in the next section. However, one of the top best practices to keep your platform secure over time is to keep on top of updating and patching your version of Kubernetes.

The Kubernetes upstream community releases three major versions per calendar year, and then there are smaller security updates and patches, usually made available as minor version releases throughout the year. It's very important to keep up with the security and patch minor-version releases while the major upgrades are in the works. Equally, as a best practice, it's important to follow the upgrade cadence for major releases as falling too far behind will mean that new releases will no longer be supported from a security-fix standpoint. Likewise, it will make the upgrade cycles more complex to manage as they will require smaller incremental upgrades. Most vendor-supported Kubernetes distributions automatically apply multiple incremental upgrades to get to a target version. They also often support one-click upgrades and zero-downtime upgrades. Community distributions, on the other hand, are not fully automated and require manual input to complete the upgrades.

It is therefore the prerogative of the enterprise to plan their resource capacities appropriately, especially when it comes to adhering to the upgrade cadence. At no point should a lack of resources (be they system or human) be quoted as a reason for not being able to upgrade or patch the cluster in a timely manner. Future scaling requirements, agility, complexity, and the ever-changing technology landscape should all be taken into due consideration when making this critical choice.

The application layer

The application layer is the top of our stack as we traverse to the edge. This is the most difficult layer to apply standardization and enforcement to as far as security best practices are concerned. The primary reason for this is the different kinds of applications out there. Even in the example that we use in this chapter, we could have a browser-based, web-based, or mobile-based application. On the platform side, we could have applications that are built on a database or that are container-based.

A very important aspect of application security encompasses the protection of the data that the application accesses underneath. We will examine the question of protecting our data in detail in the *Protection of data* section of this chapter.

Application security encompasses the security best practices of the entire **software delivery life cycle** (**SDLC**) pipeline. This pipeline is usually referred to as the DevOps pipeline.

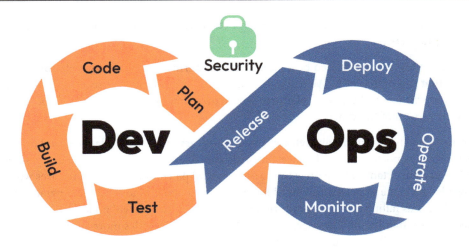

Figure 6.7 – Boundaries of responsibility

When the DevOps pipeline is integrated with security practices, it's referred to as a DevSecOps CI/CD pipeline. Some of the integrated practices that may require additional tooling include the following:

- Static code analysis (A method of analyzing the computer program code without executing the program)
- Vulnerability scanning of containers (Compare the contents of a container against a database of known security vulnerabilities)
- Threat modeling (A form of assessment that models possible security threats and defenses against them)
- Threat intelligence (Understand the data collected to derive intelligence on security attacks)
- Compliance-required policy enforcement (Automatically enforce security policies required for compliance, we will see this aspect more in detail, later in this chapter)
- Container image signature verification (Digital signature verification of the container images to verify authenticity)

Securing the artifacts throughout the delivery pipeline plays a major role in enforcing security for the application in the production environment. Protecting the software production chain itself by deploying the application artifacts as immutable containers in a Kubernetes environment is an additional security practice to prevent threats and vulnerabilities arising from CI/CD chain tools.

So far in this chapter, we've understood the components of hybrid cloud and applied them to an illustrated example of a simple hybrid cloud application that included a legacy database in the on-prem data center, a data pipeline, an application hosted inside a container on a public cloud, and an edge Kubernetes cluster that transmits processed data from IoT devices from an edge location. Using this example, we traversed through the entire application stack of infrastructure, platform, and the end-user application and understood the security considerations at each layer. In the next section, we will plunge into security best practices.

Configuring security within a hybrid cloud environment

Hybrid cloud security configuration involves multiple components across the entire application stack. Let's now go a level deeper into what it actually means to configure security within a hybrid cloud environment. It is again important to reiterate the fact that security is always a shared responsibility, and so we will highlight who is responsible for the security of a given component when relevant.

Configuring identity

Identity spans all components of security. We mentioned previously the concept of zero trust and the extension of trust boundaries across multiple systems. Hence, it's important for all components to always present an identity. Identity impacts many personas (user roles such as developers, sys admins, etc.) in the world of hybrid cloud and it's important to apply the security principles consistently and appropriately across all the entities while delivering access, services, and data. Upholding these principles while adhering to the compliance and regulatory requirements of the business is the key to architecting a secure hybrid cloud solution. We will look at this in detail in the *Compliance and governance* section.

Identity serves as the entry point into all aspects of security. There are multiple personas in the world of hybrid cloud and in this section, we will consider all of them through the lens of security. Each persona can be seen as a group of users and each group usually has a set of access controls and permissions. This process of granting permissions to entities to allow them to access, read, or write data and changes in the system is called **Identity and Access Management (IAM)**. These permissions are bundled into roles that define the level of access and can be pre-defined or custom-made depending on the granularity of access rights required. Identity itself consists of two parts: **Authentication (AuthN)** and **Authorization (AuthZ)**, which essentially mean *who is* allowed to do *what*, respectively:

- **AuthN** determines who you are, and comes from an identity platform:

  - End users (human)

  - Federated login through social platforms

  - Service accounts for different systems and cloud providers (GCP – Service Accounts, Azure – Service Principals, AWS – IAM User)

  - API key credentials (usually associated with a user or service account)

  - Certificates for SSL/TLS

- **AuthZ** determines what you can do, and is governed by the following:

  - Roles (admin, user, etc.)

  - Groups (Dev, QA, SRE, Ops)

  - **Access control lists** (ACLs)

  - Context

  - Location (e.g., restricted country)

To provide in-depth security analysis, every layer of the architecture is likely to need some notion of identity in order to be able to create logs, record metrics, and access data.

Now let's look at some of the best practices when it comes to IAM:

- Implement single sign-on or federate identity through a single identity platform

- Centralize AuthZ using a single repository

- Follow security best practices regarding the super admin account

- Automate the user life cycle management processes

- Administer user accounts, service accounts, and groups programmatically

- Use the principle of least privilege while assigning permissions and consider creating custom roles in public clouds for granular access control

- Constantly monitor and audit the permissions granted against the usage and activity of the respective user/service account or group

- Automate the rotation of user-managed service account keys

- Challenge requests for service accounts to analyze and understand whether what is requested is actually necessary

- Always consider using short-lived credentials

- Secure all API access with credentials

- Create password policies for user accounts and configure 2FA

The most important aspect of hybrid cloud security is to properly design and manage trust and security boundaries across multiple systems. Traditionally, enterprises had to master the art of sequestering their network perimeter to restrict access across components of the application, all the systems running within their data center. Now with hybrid cloud, we have platform systems running in one or more public cloud environments. This engenders the need to expand the network trust boundary well beyond the logical network within a data center. Even though enterprises have now become customers of public cloud providers, they should retain the role of the trust issuer for greater control.

Securing the Kubernetes platform

In this section, we will focus on the security of the infrastructure, platform, and workloads in Kubernetes. The following diagram shows the layers in the Kubernetes platform that need to be secured:

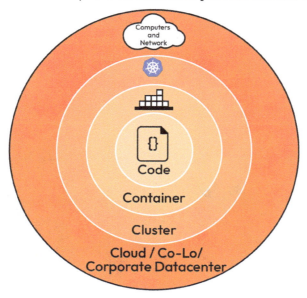

Figure 6.8 – Layers within the Kubernetes platform

We will start going up the layers, starting with securing the hardware used to create the cluster and considering the best practices to create a secure Kubernetes cluster. We will then cover network and data security (encryption) in the subsequent sections.

Infrastructure security

Kubernetes as a platform does a good job of providing distributed computing while abstracting the infrastructure. Here are some of best practice recommendations to keep this layer secure:

- Use of optimized operating systems suited just to running containers to limit the attack surface

- Limit platform access to the underlying operating system (i.e., limit syscalls and filepath access)

- Integrate the crypto validation capabilities (discussed in the previous section) offered by cloud providers during node bootstrap (i.e., when joining new nodes to the cluster)

The best way to keep your cluster secure is to make sure you are running the latest version of Kubernetes. When enterprises start running mission-critical workloads on Kubernetes, upgrading without significant downtime is an explicit requirement. It is important to note that choosing a vendor distribution of Kubernetes shifts responsibility for the aforementioned steps to the vendor and provides a great uplift with seamless upgrade, maintenance, and administration of the clusters usually referred to as Day 2 operations.

Platform security

A key security control for Kubernetes clusters is strong **Role-Based Access Control (RBAC)** to ensure both the cluster users and the workloads that run on the cluster have the appropriate level of access required to execute their roles. Access to the Kubernetes datastore (etcd) in particular should be limited to the control plane only, and etcd should only be accessed over TLS. It is a good practice to encrypt all storage at rest; etcd is no exception to this, as it's the brain of your cluster.

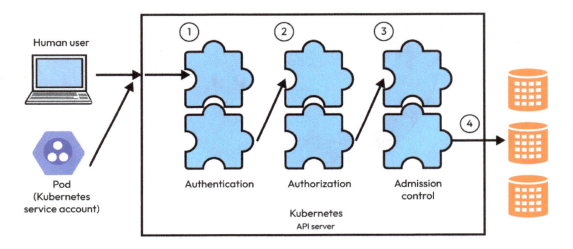

Figure 6.9 – Security path of a request to a container

The preceding diagram shows the security path of a request to a container running in the cluster. Any request to the API server, whether it's from a human user or a service account, has to go through authentication and authorization, followed by checks from the admission controller before any change of state can be made to the entity within the Kubernetes cluster.

API authentication

All clusters must be set up with authentication mechanisms and all API clients must be properly authenticated before the API server will process any request. The lower-level test environments such as dev, which sometimes has just a single user, could use a static bearer token approach, while the higher, larger, and more secure environment clusters could be integrated with existing OIDC or LDAP servers, which allow for more granular access control based on groups.

API authorization

Once authenticated, all API calls are expected to pass an authorization check. For this, we will apply all the best practices discussed in the *Configuring identity* section of this chapter. Kubernetes ships a comprehensive RBAC stack that helps to match an incoming request's user or group to a set of permissions or operations that are bundled as roles. These permissions are a combination of verbs such as *get*, *create*, or *delete* with Kubernetes resources such as pods, services, or nodes with scopes that either limit them to a namespace or have the permissions implemented cluster-wide.

The same principles of authentication and authorization must be applied to the kubelet as well, since by default the kubelet allows unauthenticated access to the API server. Every time a new node is added to the cluster, a process called bootstrap needs to be followed where the kubelet process running on the nodes establishes communication with the control plane securely, such as by using TLS certificates.

As part of setting appropriate privileges for workloads, it's important to define appropriate platform security policies as part of the pod security standards. The policies themselves are cumulative and range from highly restrictive to highly privileged. Since Kubernetes is a declarative system by design, the entire security posture of the platform can be represented using a set of configuration files. These files include both the core configuration of the cluster and the security policies associated with the environments served by these clusters. Storing these configurations in a version control repository and managing these files similarly to the source code of an application is an example of *config as code*. We use these principles and practices to deliver a GitOps way of platform administration.

Workload security

When it comes to implementing a zero-trust strategy, auditability and automation are two key principles. The recommended security best practices for workloads also revolve around these key principles.

Kubernetes provides **admission controllers** that help to perform validations on a request coming into the API server (essentially a change to the configuration file since Kubernetes is a declarative system) prior to the persistence of the change to the Kubernetes object storage. This helps us to be proactive with our security-related pre-checks. A good example of this use case is checking for the signature of images or for the presence of explicit metadata before allowing the deployment of those images in the cluster. Another example of the admission controller is pod security admission, which helps to enforce the pod security standards.

Pod security admission places requirements on the pod's security context and other fields that help to enforce the pod security standards. These standards fall under one of the following:

- Privileged (an unrestricted policy, providing the widest possible permissions)
- Baseline (a minimally restrictive policy that prevents known privilege escalations)
- Restricted (heavily restrictive, following pod-hardening best practices)

In the enforcement of pod security standards, the pod security context plays a big role. It defines the privilege and access control settings for a pod or container and is part of the pod configuration definition.

The application container images are actually treated as infrastructure as well. These images are subject to security vulnerabilities and hence it's important to efficiently manage their inventory. The following picture shows the typical life cycle of vulnerability and compliance management:

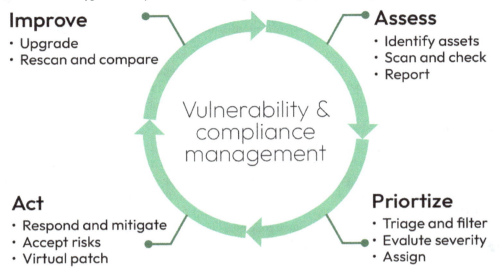

Improve
- Upgrade
- Rescan and compare

Assess
- Identify assets
- Scan and check
- Report

Vulnerability & compliance management

Act
- Respond and mitigate
- Accept risks
- Virtual patch

Priortize
- Triage and filter
- Evaluate severity
- Assign

Figure 6.10 – Life cycle of vulnerability management

Vulnerability threat analysis must not be restricted to just scanning the container images in the registry. Scanning for vulnerabilities must be extended to those containers running in production – a process also known as dynamic threat analysis. The vulnerabilities found from this scan can be categorized into different levels of criticality. Whether an image is found to be compliant or not depends on the enterprise's own defined response to these levels of criticality. Checking image signatures and scanning for vulnerabilities can be automated within your software delivery life cycle.

If a container is not built correctly, then there is a chance that it can run as privileged. This would give the container unnecessary access to the operating system and introduce a security vulnerability. It's important to make sure the platform is configured in such a way that the principle of least privilege is followed, so the container gets only the necessary level of access required to fulfill its tasks. Container developers should also try to use container runtime classes where appropriate, as they provide a stronger level of isolation.

Code security

In addition to the best practices discussed around Kubernetes security, here are a few best practice recommendations for your code:

- If any application service needs to communicate over TCP, make sure to only access it over TLS. It is also a good practice to encrypt everything in transit.

- Expose only those ports that are absolutely necessary for communication or gathering metrics.

- If the application uses any third-party libraries, it's a good practice to scan them for known security vulnerabilities.

- Integrate static code analysis and dynamic probing into your DevSecOps processes.

- It is also a good practice to include a software bill of materials as part of the artifacts in the software supply chain.

In this section, we went a level deeper to understand the role of identity in securing our hybrid cloud environment, as well as examining the best practices around IAM. This defines how any two entities within a hybrid cloud environment can communicate with each other (i.e., with users or with other systems). We also looked at how to secure the Kubernetes platform at the infrastructure, platform, and application (workload) layers.

Next, we are going to look at how to secure the network aspects of a hybrid cloud environment. This topic is very important as it plays a crucial role in extending trust boundaries. Also, given there are so many moving parts in the hybrid cloud, a secure network layer improves the security posture of the entire architecture.

Configuring network security

No matter whether a public cloud, a private cloud, or the edge, there are always two components to networking. One is the physical network, while the other is the software-defined network, an overlay network on top of the physical network. As we discussed previously, in the on-prem and the edge parts of hybrid cloud, the onus of responsibility for security is on the enterprise, while the public cloud provider guarantees a basic level of network security and provides networking services and capabilities to integrate security into the design. The following diagram shows the network boundaries that are typically implemented while creating the network architecture:

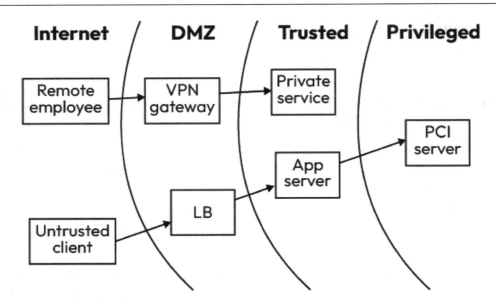

Figure 6.11 – Network boundaries

The preceding diagram illustrates two important considerations to bear in mind when designing networks. Both are paramount in a hybrid cloud architecture:

- Configuring network perimeter controls
- Configuring network segmentation

These can be implemented by using network firewall rules to create air-gapped networks that restrict internet access completely and create a DMZ (or demilitarized zone), which is a perimeter network that adds a layer of security around the internal network for all untrusted traffic.

Before exploring these considerations more deeply, it's important to determine the entry point into our application. In other words, where does the public endpoint that provides access to our application reside? The breakdown of the network design depends on this location. This endpoint can either be in one of the public cloud's public address spaces, or within the data center or private cloud of the enterprise. For the sake of consistency, we are going to pivot back to the same example used in the *Different components of security in hybrid cloud* section. Based on the architecture in the example, we are going to assume that our public endpoint resides in one of the public cloud providers.

Assuming the endpoint resides in the public cloud provider, it will usually be exposed by a secure load balancer. Cloud providers have their own web application firewalls that protect this endpoint against **Distributed Denial-of-Service (DDoS)** attacks. Enterprises also have the flexibility to choose a cloud-agnostic WAF solution to keep design and usability consistent across their on-prem systems and multiple cloud environments, depending on their strategy. The load balancer can either be on Layer 7 or Layer 4, depending on the requirements of the application.

It's important to secure traffic to meet the security and compliance objectives. A best practice recommendation is to use the **Secure Sockets Layer** (**SSL**) or **Transport Layer Security** (**TLS**) protocol to secure data transported through the network. Also, the endpoint exposed will have a DNS name associated with it. Another best practice recommendation is to use **Domain Name Security Extensions** (**DNSSEC**) to protect your DNS data. DNSSEC provides cryptographic verification of data authenticity from the source and offers integrity protection (guarantees against data modification in transit) to the DNS protocol.

Now, we move into the **virtual private cloud** (if the cloud is GCP or AWS) and virtual networks (if the cloud is Azure) that host the application and the data pipeline. This is where we have two other networks integrated with the core VPC – we have the edge network hosting our Kubernetes cluster and we have our legacy database in our enterprise data center. Here is where the network perimeter controls become very important. Firewall rules play a very important role in controlling the ingress and egress traffic through the entire network. In addition to the firewall rules, we can also segment the network, classifying the resultant sub-networks as those that are publicly accessible and those that are completely private. We also have the concept of routes in our network, which define the path taken by the network packets from the source address to the destination address. Using a combination of network perimeter controls and network segmentation, we are able to provide network isolation when we have multiple tiers in our application. The following are a couple of common examples of network segmentation and isolation:

- The hub-and-spoke networking model, where all external network connections exist within a certain VPC that acts as the perimeter space for connections
- Different layers of the application such as the web tier, app tier, and database tier are in different subnetworks

These segmentation and isolation practices when employed in the design help to minimize the blast radius when security breaches occur or any component of the architecture is compromised.

Before we move to securely connecting external networks to our VPC, let's explore in detail some networking best practices for our Kubernetes clusters.

Kubernetes network security

Since the Kubernetes platform is a distributed system that brings together a set of individual machines to form a *cluster*, networking plays a very important role in the architecture. Each machine represents a node in the cluster, and this machine can be either a virtual machine running on a hypervisor (VMware/KVM/Hyper-V) or a virtual machine instance running a cloud provider's infrastructure or a bare-metal server. The specific container runtime used on each machine (node) that gets deployed during the cluster installation has its own implications for security as well. The most common container runtimes use **Container Network Interface** (**CNI**) plugins to manage their network and security capabilities. In this section, we are going to look at some best practices and security controls for creating and running Kubernetes clusters.

Kubernetes is entirely API-driven and hence all access to the cluster is done through the API server that is part of the Kubernetes control plane. Here is a list of best practices associated with controlling access to the API server through a network ACL for network security:

- Always disallow public access to control plane nodes

- Allow access to the API server only from authorized networks

- Disallow public access to the nodes of the cluster

- Nodes should always be configured to accept connections only from the master control plane network

All API communication is expected to be encrypted in the cluster using TLS as part of building a zero-trust architecture. However, if this zero trust needs to be extended to the services created by the application, then the need for a service mesh such as Istio arises. Service meshes are out of the scope of this book, but we will quickly explain the use of a service mesh from a security standpoint. Besides being able to connect, manage, and monitor various services running in the cluster as part of your application, a service mesh helps with the implementation of network security via the following:

- Encryption in transit

- Zero trust based on mTLS authentication (where appropriate)

- Context-aware access control

- Policy-driven security for services

However, if the application architecture is not microservice-based and only requires basic network security for its workloads, Kubernetes also offers networking plugins that can be applied on top of the **Container Network Interface (CNI)**. This helps to control the flow of traffic at the IP address or port levels. These are called **network policies** and are application-centric, essentially defining the entities with which a pod can communicate.

By default, Kubernetes networking allows communication between pods even when they are distributed across nodes, there is no brokering of port numbers, and all pod IPs are cluster-scoped.

Hence, network policies are of significance when a cluster is used for multi-tenancy. Network policies help to institute policy-based isolation. This also provides a key functionality for implementing specific compliance requirements, such as PCI. Some examples of network policies are the following:

- Default deny all ingress traffic

- Default deny all namespace-to-namespace communication

- Allow all ingress and egress traffic

- Default deny or ingress and egress traffic

Establishing private connectivity

The network perimeter controls and network segmentation discussed at the beginning of this section are applicable to each of the individual networks that are part of our example architecture. For us to extend the trust boundary for both the on-prem data center and the edge network, we need to establish private connectivity between the respective entities and the public cloud. There are two ways that we can extend the RFC 1918 private address space of our VPC in the public cloud:

- Direct Interconnect or Partner Interconnect from the data center to the public cloud or the edge network
- The VPN tunnel between the data center or the edge network

The choice really depends on the bandwidth requirements of the connection. VPN bandwidth is usually much lower than interconnect.

Protection of data

In the previous section, we looked at the network layer to understand the best practices along the journey of a request from the end user to the application. We also looked at other layers of the application stack and made sure each component of the hybrid cloud infrastructure was secure against any attack or infiltration. In this section, we are going to look at how we can protect the data that exists in all the components of the architecture. As with all other components, responsibility for securing data is also shared.

In our hybrid cloud infrastructure, there are many kinds of data that we need to protect. When we refer to *data security*, it's not about just protecting the application data. The data in our architecture can be of many types:

- Infrastructure and systems information
- Network logs
- Platform logs
- Monitoring and metrics information
- Application logs
- Application data
- Transactional data
- Archived data
- Analytical data

Application data, as in the preceding list, can be broken down into three categories:

- Data in use by applications; that is, being processed in the CPU or memory. Encrypting this in-use data is the core idea behind confident computing.
- Data that is in transit across our network.
- Data at rest; that is, stored anywhere in our hybrid cloud infrastructure.

The work to protect this data is made up of two main areas:

- Data security and privacy
- Data governance

So, what is the difference between data security and data governance? Data security is all the controls that are put in place to protect data from unauthorized access. Data governance defines the policies and procedures employed for maintaining security and compliance. In this section, we will cover data security. We will look at data governance in the later *Compliance and governance* section.

The best way to ensure robust hybrid cloud data security is to secure and control different data types listed earlier using proper IAM. In the context of hybrid cloud, this needs to be applied across multiple networks and infrastructure. Since logs and system data are dispersed across many entities, the best practice is to unify all this data in a centralized location to monitor and analyze it. We will cover this in more detail in the *Securing hybrid cloud operations* section. IAM best practices also have a significant impact on data governance related to enterprise compliance requirements.

As mentioned earlier, besides protecting application data using IAM and the principle of least privilege, we need to employ other ways to protect data in transit and at rest. In the *Configuring network security* section, we discussed protecting data in transit using the mTLS protocol. In this section, we are going to discuss protecting data at rest. Most public cloud providers offer this feature by default for the data used by cloud data services and for all systems and log data. We are going to look at the mechanics of data protection and control in detail. Let's pivot back to our example and focus on the data pipeline. The following are the various stages that are typically associated with a data pipeline:

- Ingestion/collection
- Processing
- Storing
- Analyzing

Encrypting the data when it's stored ensures its protection at rest. This is a key requirement whenever data is stored in any system. Data is encrypted using encryption keys. The creation and management of encryption keys defines the level of control required by an enterprise based on their data governance and compliance requirements. The following is a list of different options for managing encryption keys:

- Cloud provider-managed encryption keys

- Enterprise encryption keys (both created and managed)

- Keys managed by external or cloud-agnostic key managers

As discussed in the previous chapters, the state of the Kubernetes platform is stored in etcd. It's important to secure that data at rest using one of the preceding encryption options. Let's look at securing the data associated with Kubernetes.

Kubernetes data security

As with protecting all other data at rest using encryption, it's a good practice to encrypt the data stored in etcd (which holds the state of the entire cluster). Kubernetes also provides *secrets* to enhance the security capabilities of your application. A secret is an object that contains sensitive information, such as a password. These get stored in etcd as well, so there is all the more reason to encrypt etcd data at rest. Since secrets help with the protection of data, the following are a few best-practice recommendations for secure usage:

- Enable encryption at rest

- Configure RBAC

- Restrict access to workload-specific containers

- Use external secret store providers

In summary, we have seen the various best practices for securing data in our hybrid cloud architecture. Using a combination of good IAM and data encryption practices, we can prevent unauthorized access to the application data. We also listed other types of data generated by our architecture and how those can be secured as well. In the next section, we will focus on securing the operations aspect of hybrid cloud.

Securing hybrid cloud operations

So far in this chapter, we have looked at best practices to prevent access to unauthorized data in our hybrid cloud architecture. Since the hybrid cloud architecture is dispersed across different infrastructures, operations become very challenging. Beyond application data, a lot of systems data is also generated by our hybrid cloud infrastructure and platform. Besides securing that data using IAM best practices, there are other responsibilities for us to be aware of (some of which we listed earlier) as part of security around operations.

Tasks as part of building and deploying secure infrastructure and applications include the following:

- Automate creation, hardening, and maintenance of base VM images (nodes) and base container images (applications)

- Automate security scanning for common vulnerabilities and exposure through CI/CD as part of the DevSecOps life cycle
- Automate the detection of potentially dangerous behavior at runtime

Configuring centralized logging and monitoring allows us to be proactive in tackling security issues and efficiently manage the overall security posture of the enterprise. It's important to constantly monitor and analyze network logs, including firewall logs, network flow logs, and packet mirroring, especially within sensitive and regulated industries. Analyzing and monitoring cloud audit logs and data access logs should also be a routine part of your security operations.

Compliance and governance

We have systematically looked at securing every component of the hybrid cloud architecture so far. As discussed earlier in the *Protection of data* section, data governance is the set of rules, practices, and processes put in place to provide oversight of data security and help the business adhere to its compliance requirements. Besides preventing unauthorized access to infrastructure and data, all businesses have the responsibility to protect the data privacy of their consumers. This is especially true in regulated industries such as public sectors, finance, and healthcare. Failure to do so comes at a huge price for a business.

Let's explain compliance using a framework to define some key components of it. We discussed some of these components in earlier sections of this chapter:

- **Governance**: A set of rules that defines the access and control of all resources and data in the hybrid cloud environment.

- **Change control**: Following the principle of least privilege, no system or user should have more privileges than required to fulfill their tasks. Changes made to any entity in the hybrid cloud environment should be controlled by a proper process as part of governance.

- **Continuous monitoring**: As mentioned earlier, due to the dispersed nature of the hybrid cloud system, monitoring and logging become very important pieces of the compliance puzzle.

- **Reporting**: Proper reports help to progress projects within the swim lanes and ensure adherence to all compliance requirements. Reports are useful to provide material evidence if there are any security incidents and compliance is called into question.

All businesses are bound by sets of regulations, depending on their type or vertical, defined by the relevant legislative body. For example, organizations looking to do business with US federal agencies and who choose to employ a hybrid cloud architecture can only choose a cloud provider that is FedRAMP-certified. Similarly, the European Union's GDPR outlines rules covering data residency for data protection. Adhering to these rules and regulations is required for an enterprise to be in compliance. This is usually certified by an external body. In simple terms, what does ensuring compliance mean in the hybrid cloud world?

- Implementing rules defined for compute, data, and network

- Compliance is based on the shared responsibility model

- Configuring security within the hybrid cloud environment

- Provide access using the principle of least privilege

In addition to data security guidelines, data governance also defines how long this secure data needs to be stored for. This is called **data retention** and is part of the data life cycle. This can vary based on the compliance requirements of the enterprise. Data governance also helps to define the classification of data. Data classification might determine the level of encryption and control required to be applied to that data. Data governance might also require data masking or data deidentification on specific attributes of data to enforce data privacy.

The challenge with hybrid cloud is not about one specific compliance or governance requirement, but arises due to the dispersed nature of the architecture across many systems and networks, so the overall level of risk is much higher. It's often tedious and error-prone if all these checks have to be done manually. Choosing tools and solutions that can be used consistently across all parts of the hybrid cloud environment helps solve at least part of the problem. In addition to adhering to all the best practices mentioned for each layer and within each component, security can be implemented at scale only if all the processes, practices, and checks are implemented in an automated manner. Tools that provide a single-pane-of-glass view of the entire architecture greatly help enterprises to counter the threats out there and be proactive with their security measures.

Summary

All layers in our application stack have data associated with them that necessitates a consideration of security. Responsibility for security lies with the enterprise or the cloud provider in differing proportions depending on the infrastructure footprint. Any architecture can be approved only after a detailed end-to-end security analysis.

Throughout this chapter, we focused on the different components of hybrid cloud across the entire application stack to understand the security implications of each layer from an architecture standpoint. To further understand the security aspects objectively, we looked at an example of an application that used a hybrid cloud architecture. This is just one of many architecture patterns that exist for hybrid cloud, but the fundamentals remain the same across other patterns.

The principles of security discussed weren't something new, but are important for enterprises to acknowledge the danger associated with architectures that do not adhere to the recommended best practices. In the next chapter, we will look at the economics of hybrid cloud and the steps to run a hybrid cloud infrastructure in the most cost-effective manner. However, it's paramount to stress that implementing the steps to optimize costs should not come at the expense of implementing a security best practice.

7
Hybrid Cloud Best Practices

Before we get into the best practices of hybrid cloud, let's take a minute to understand how we got here. There is a proliferation of cloud computing, and traditional IT practices are being forced to evolve. As we embrace the growth of cloud computing, it's impossible not to acknowledge the increase in complexity and its predominance in different facets of IT. Besides organizations creating their own private infrastructure, they also have access to the major public cloud providers of Amazon, Microsoft, and Google. As we embark into the world of hybrid cloud, architects and engineers have to deal with the following challenges:

- Different connections across different systems

- Integrations

- Application portability between clouds and data centers

- Optimization of resource utilization and consumption

- Infrastructure and application life cycle orchestration

- Storage

These challenges are by no means comprehensive as we dedicated an entire chapter to the security of the hybrid cloud. In the previous chapter, we looked at the different components of hybrid cloud architecture and how we can secure them. Hybrid cloud architecture is complex since it is dispersed across different entities (on-premises, data centers, public clouds, edge locations, etc.). Hence, we took a holistic approach when it came to security and tried to recommend best practices at each layer of the application, including the network, infrastructure, and platform.

Together, we understood how we can secure not just the application running in the hybrid cloud but also the data that is being stored and accessed throughout the architecture. In this chapter, we are going to learn about best practices for rolling out a hybrid cloud. The concepts discussed here will help to increase efficiency and cost-effectiveness, using techniques and guidelines that make the most of your hybrid cloud investments.

We will cover the following topics to understand the overall best practices around implementing the appropriate architecture for the appropriate use case. The first step is to understand the important guidelines for a cloud strategy. Then, we will go into the details of developing the architecture based on the guidelines we have learned and understand the various options that are available to deploy and maintain the architecture. We will also provide insights into the economics of hybrid cloud and discuss the various factors that influence the overall cost of the implementation, and finally, wrap up with the challenges and pitfalls of hybrid cloud that are associated with this implementation. These are the topics we will cover:

- Guidelines for a cloud strategy

- Understanding various architectural considerations

- Development and deployment of hybrid cloud

- What do we mean by the economics of hybrid cloud?

- Various pitfalls of hybrid cloud

Guidelines for a cloud strategy

The first step of defining a cloud strategy is aligning the IT goals and objectives with the key business objectives of the organization. This alignment requires a strong foundation that involves a streamlined IT process. Let's analyze the need for a cloud strategy in the context of business use cases:

- **The application requires auto-scaling**: If the business is not able to provide a consistent requirement for resources, IT will turn to more robust ways of optimizing resource utilization. Elastic scale-up and scale-down are usually termed *auto-scaling,* which is a good functionality. The cloud provides this option and bleeds into the next business problem.

- **Capital expenditure (CapEx)** versus **operational expenditure (OpEx)**: As an IT organization, do I invest in more resources or do I pay only for resources used based on a consumption or pay-as-you-go model? Which parts of my application stack can I consume as **software as a service (SaaS)** or use sub-contracted services that minimize the requirements for hardware? Can I repurpose or procure new hardware and offer it at a scale matching my business expectations?

- That brings us to the most important advantage a public cloud offers: speed, agility, and effectively meeting the time-to-market pressure of the business.

- Data proximity is required for regulatory compliance and governance.

The guidelines listed here just provide a structural way of assembling the key ingredients. These considerations lead to making appropriate architectural determinations. The use cases then become the eventual starting point for the architecture of the hybrid cloud.

Understanding various architectural considerations

Before going into the architectural considerations for a hybrid cloud, it is first important to understand the problems or challenges that we are trying to fix using a hybrid cloud implementation. As with all architectures, the hybrid cloud has its own pitfalls as well. We will look at this in more detail in the *Various pitfalls of hybrid cloud* section of this chapter.

Let's keep the cloud strategy highlighted in the previous section as a good background and understand the various architectural considerations:

- Portability and manageability help to avoid vendor lock-in

- Data residency and regulatory requirements

- Application requirements for regional and global redundancy

- Efficient software delivery supply chain processes across the entire infrastructure footprint

The following diagram is the same illustration that we used in *Chapter 6* and shows a simple web application running inside a Kubernetes cluster that can be accessed by users from a desktop, laptop, or mobile device. The application has data stored in a cloud-native database that gets populated through a data pipeline. It receives data from an edge cluster that is running Kubernetes, which preprocesses data from IoT devices, and also a legacy data source that is running in an on-premises data center sending data in batches.

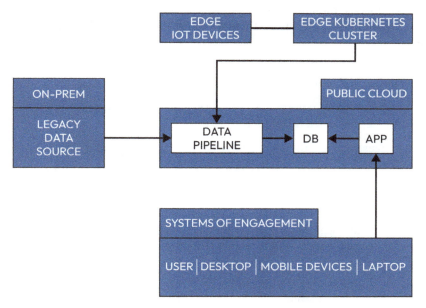

Figure 7.1 – Sample hybrid cloud architecture from Chapter 6

Given this example, here are a few scenarios where it makes sense to use a hybrid cloud architecture:

- You are unsure about the data analytics and data management tools.

- In a multi-cloud world where consistency is the key, **Platform as a Service** (**PaaS**) providers usually offer automated deployment of data analytics and data management solutions that are cloud-native but also cloud-agnostic.

- The cloud provides the ability to fail fast and fail safe.

- Engineering teams do not have to go through IT service management practices and processes to see whether this is indeed a viable option by trying to procure all hardware in-house.

- The maintenance of the hardware and VMs becomes a burden, trying to ensure that the OS is constantly updated and patched, when this is natively taken care of by the cloud provider.

- You are unable to accurately determine the demand for and scale of the hardware.

- There is periodic demand for the solution to provide elastic scalability.

- Redundant hardware is provisioned for HA architecture.

- It suits enterprises who choose a cloud-first approach for new workloads while leaving the legacy systems in their data center.

- Not all components of the stack can be modernized.

- There is a shortage of resources to manage and scale new application's hardware and software requirements on premises. These are classic use cases for bursting into the cloud for additional capacity.

- Latency-sensitive workloads and workloads that require high availability across cloud providers fall under a multi-cloud hybrid cloud strategy.

- Given cloud providers are more globally dispersed than enterprise data centers, certain frontend components may be hosted on the cloud for greater access and outreach.

- Cloud services provide the quick start capability and remove the burden of day 2 operations completely.

- Many times, public clouds provide better connectivity to edge locations.

- Besides all these, different applications have different demands across the entire stack, including SLAs on availability and performance. It's very difficult for the enterprise to meet each of these requirements in the most cost-effective and efficient way.

- Workloads require infrastructure for high-performance computing (data scientists).

Once we know the reason for choosing a hybrid cloud, it's important to clearly understand the hybrid cloud architectural design principles. As discussed earlier, hybrid cloud introduces new variables, such as extending the network trust boundary, and consistent security principles across all layers, and there is also a dependency on external entities to guarantee service availability. As far as the application is concerned, the end user should have a seamless experience, immaterial of where the component of the application is running.

Here are some of the design principles that can be applied throughout our architecture:

- A consistent way to operate and manage both the on-premises as well as the public, private cloud, and edge infrastructures

- Make your applications cloud-native or container-based so that they are infrastructure-agnostic and provide flexibility to run them either on premises or on the cloud with ease

- Use APIs for the creation and management of infrastructures so all components of the application are maintained in an automated manner

Building further on these design principles, an in-depth architecture planning also includes considerations of the following aspects:

- **Capacity planning**: Even though we have a public cloud in our hybrid architecture, the available capacity is not limitless. Whether the use case is all on-premises and the public cloud is used for bursting, or we have some key components of our application running on the public cloud, it's important to work very closely with the business and plan capacity allocation appropriately. Adding additional capacity to the on-premises infrastructure or the private cloud should be carefully weighed against the pros and cons of CapEx versus OpEx. It is also prudent to reuse or repurpose existing hardware through application and platform modernization initiatives, which invariably frees up more capacity in-house. The following figure shows a hybrid cloud architecture where additional capacity is utilized through the public cloud. This is a common design pattern for cloud burst scenarios:

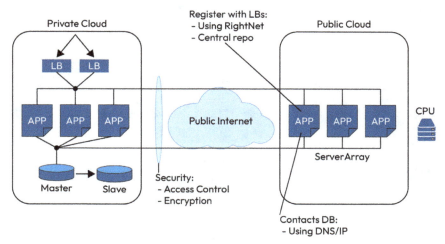

Figure 7.2 – Adding additional capacity using a public cloud

Here are some of the best practice recommendations for capacity planning. It is important to establish a good strategy on how to use the cloud to expand capacity. Setting up clear guidelines and guardrails helps to quell unwanted requests to increase capacity made by engineers for their services. One such guardrail is showback and chargeback applied to the business unit's resource utilization.

This should be followed by a clear process laid out to increase capacity when required. Publishing this to all teams, along with clearly defined SLAs, helps teams to factor this into their project plan. It is also important not to overcommit production environments. Instead, the architecture should be optimized to the respective cloud providers' strengths and make use of the auto scale-up and down capabilities provided by the cloud provider. Finally, it all comes down to how efficiently the workload is architected. The Kubernetes platform, as we have seen in this book, is a great example of a cloud-native, cloud-agnostic, distributed computing platform that helps optimize the resource utilization of workloads:

- **Service-level objectives (SLOs) against Service level agreements (SLAs)**: This aspect of the hybrid cloud will test the resiliency of the architecture. The application components are spread across different infrastructure entities. Amid all the variable factors and complexities, the core SLA of the application has to be maintained. It is a factor of cumulative SLAs of the public cloud and private/on-premises infrastructure, edge (if applicable), hybrid network connectivity, and so on. It's imperative to define SLOs at each level when designing for availability, and it must be also factored in while building redundancies into the architecture and weighed against the business continuity guidelines for the application. Every service should have an availability threshold defined and appropriate tools to monitor and collect metrics, and ways to collate and observe them on dashboards. The following figure shows the extension of the hybrid cloud to edge and IoT devices:

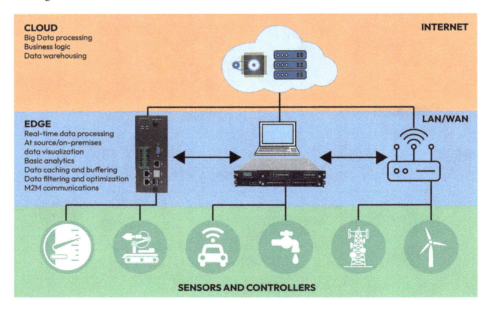

Figure 7.3 – Hybrid cloud with edge and IoT devices

- **Governance around spending**: Governance around spending is a factor in capacity planning. With the availability of "infinite" infrastructure capacity, if the usage is not gated, then there is a possibility of the cost going out of control. A good architecture will define the limits for infrastructure for all components of the application. Appropriate alerts must be created in the respective monitoring tools, so the performance of the application is not compromised. It's always a fine balance between getting the design of the application right using the optimum resources without impacting performance. A simple example of this is to determine the network bandwidth required by the application and appropriately secure the bandwidth for hybrid connectivity.

In earlier chapters, we mapped out the role of Kubernetes in the hybrid cloud and how the container orchestration platform aids in both cloud-native as well as cloud-agnostic architectures. In this chapter, based on the various hybrid cloud design principles that we listed earlier and the architectural considerations, it is further clear that Kubernetes possesses all the capabilities to build a robust hybrid cloud architecture. Kubernetes is a platform that is infrastructure-agnostic. The following is a depiction of the platform's ability to seamlessly bring operational consistency across infrastructure footprints:

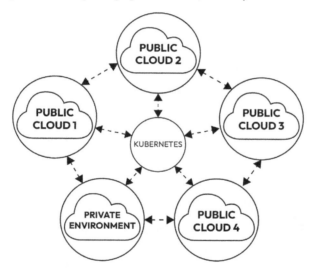

Figure 7.4 – Kubernetes for public and private clouds

Here are some advantages of using Kubernetes for hybrid cloud environments:

- Supports multiple kinds of infrastructure node types (for example, Windows and Linux) and, depending on the distribution, on different infrastructure footprints such as virtualization, private and public clouds, and edge systems

- Provides the ability to both distribute workloads across resource pools as well as isolate workloads into specific resource pools

- A full declarative system that is based on configuration

- A rich set of APIs that supports complete automation

- Provides governance for limiting and restricting resource utilization, which results in efficient capacity planning and cost management

- A rich ecosystem of third-party tools that extends many security, networking, monitoring, and management capabilities, which helps to architect robust hybrid cloud systems

- Provides excellent application portability that helps to easily move workloads from on-premises to the cloud, especially in cloud burst scenarios

- Offers excellent self-healing and auto-scaling capabilities that help to take full advantage of the elastic infrastructure capability

The following figure shows some of the value propositions of Kubernetes at different levels of platform usage:

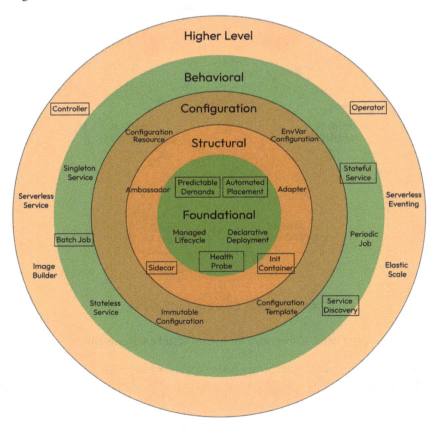

Figure 7.5 – Kubernetes value proposition for hybrid cloud

Others include resource management, enhanced container and network security, and network configuration management. Ultimately, Kubernetes is merely a container orchestration platform that brings the much-desired consistency and portability to a hybrid cloud environment.

In this section, we looked at the key design principles of hybrid cloud architecture. We also understood the capabilities of the Kubernetes platform in providing a cloud-agnostic infrastructure, while providing flexibility with application mobility and robustness with availability. In the next section, we are going to use all the concepts discussed to look at ways to develop and deploy this architecture.

Development and deployment of hybrid cloud

Once we decide on the required architectural pattern based on the problem we are trying to solve, the next step is to select the appropriate products and technologies to implement the hybrid cloud infrastructure design. The development phase includes the following:

- Compute design considerations:

  - Bare metal servers

  - Hypervisor/virtual machines

  - Identification of cloud regions

  - Hardening and certifying virtual machine images

  - OS certification and hardening

- Network design considerations and trust extension:

  - Network perimeter and firewall rules (ingress/egress)

  - Next-generation firewall appliance, if applicable

  - VPC definition and cloud network security

  - Private and public subnets

  - The **Demilitarized Zone (DMZ)** on the on-premises network

  - A **Web Application Firewall (WAF)** to protect against **Distributed Denial-of-Service (DDoS)**

  - Restricted internet connectivity to internal systems

  - Domain controller and DNS:

    - Use cloud DNS

    - Use on-premises DNS

  - Cloud load balancers versus appliance-based load balancers

  - IP address management

- Hybrid connectivity:

  - A VPN connection (with an edge location, if applicable)

  - Private connectivity with the data center

- Storage design considerations:

  - Network filesystems

  - Container storage infrastructure

  - Cloud storage:

    - The appropriate tier based on cost and performance

    - Block storage versus object storage

- Database tier considerations:

  - Legacy on-premises

  - Cloud database service

  - Cloud-native databases running within containers

- Public cloud access and billing configuration:

  - Billing accounts

  - Service accounts

  - IAM permissions

- Kubernetes architecture and cluster configuration:

  - Node configuration:

    - Type of nodes OS (Linux/Windows)

    - Node capacity (CPU/GPU/TPU)

    - Number of nodes required

    - Auto-scale threshold definitions

    - Network configurations

  - Choice of **Software Defined Network (SDN)**

  - API server configuration/kubelet configuration

  - Cluster auto-scaling properties

- Application security considerations:

  - Integration with an **Identity Platform (IDP)**

  - AuthN and AuthZ

  - CI/CD

- Architecture maintenance and support:

  - **Infrastructure-as-a-Service (IaaS)** code

  - Configuration as Code

  - Policy as Code

  - DevOps

  - Backup and **Disaster Recovery (DR)** architecture

  - Fault tolerance and HA architecture considerations

This is a very detailed list of considerations for each layer of the architecture. Within the entire list, let's focus on key items in particular (**Infrastructure as Code**, **Configuration as Code**, **Policy as Code**, and **DevOps**), as they play a very important role in the deployment of this architecture. As mentioned earlier in the best practices for hybrid cloud deployments, it's good to have them run through an automated process. This gives both reliability and repeatability given that the architecture has many moving parts and is complex in nature.

Infrastructure as Code

Traditionally, in the world of software engineering, provisioning infrastructure was a manual process. As the provisioning and automation processes evolved, automation was leveraged by creating scripts using API calls to the respective software solutions. In 2006, it further evolved when the entire infrastructure could be modeled as code and enabled the use of software life cycle management best practices. This helped automate the entire delivery life cycle of an application with tremendous improvements in efficiency. The following figure illustrates a basic workflow for IaC:

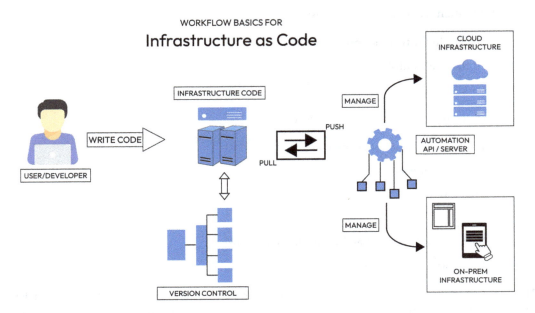

Figure 7.6 – Basic workflow for Infrastructure as Code

HashiCorp's Terraform and Red Hat's Ansible are some examples of the Infrastructure as Code language. Both are cloud-agnostic solutions, and it is a great choice to convert the deployment of your base Infrastructure as Code. They have plugins for each of the public cloud providers, as well as a private cloud implementation such as OpenStack. Modeling the entire hybrid cloud architecture and converting it into code helps to bring consistency with a repeatable process. It also helps to quickly create and destroy on-demand environments. Identical disaster recovery clones can be created accurately from the production environment code. Other advantages include the following:

- A stable environment
- Speed and agility
- Easy to maintain and access different versions
- A centralized repository for Infrastructure as Code
- Easily recreate an environment during disaster recovery
- Can be extended to software provisioning, configuration management, and application deployment
- Reduce human errors

Configuration as Code

Kubernetes provides maximum agility, flexibility, and control for maintaining and scaling hybrid cloud environments. We also know that Kubernetes is a declarative, distributed system that is completely configuration-based. Where applicable, all cloud-native and modern applications that run inside containers will pivot back to Kubernetes, the industry standard for container orchestration. The configuration YAML files are assets that define the entire configuration of the platform. Just like infrastructure, we are now able to represent our platform (if it's Kubernetes) configuration as code. The following figure shows the Git version control repository playing an important role in both continuous integration and continuous deployment, and it's the single source of truth:

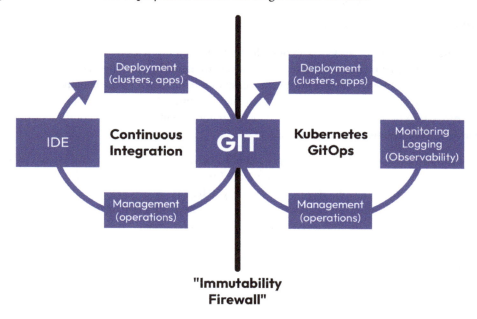

Figure 7.7 – GitOps workflow for Kubernetes

Some of the industry standard examples are as follows:

- Anthos Configuration Management from Google
- Advanced Cluster Management from Red Hat

These solutions run configuration sync operators on the Kubernetes clusters and they constantly sync with a Git repository. These operators are sensitive to any commits made to the respective repository and branch from where they derive their configuration. This helps to completely eliminate any drift and provides the platform operators with the ability to manage multiple clusters at scale, providing, at the same time, an approach to disaster recovery.

The configurations need not be restricted to just platform only but can be extended to applications as well. All Kubernetes-based application code is, again, a set of YAML files that represents the state of the application. So, it already exists as Configuration as Code. When it comes to applications, in addition to core application configuration, they also have environment configuration. If a Configuration as Code approach is adopted, it's also a good practice to separate the application code and environment configuration into separate repositories. In addition to the advantages previously mentioned for Infrastructure as Code, Configuration as Code extended to an application also has the following advantages:

- Greater security as we have a clear demarcation of the Dev and Ops boundary

- Detailed traceability for audits

- Better management of the software delivery life cycle

To further simplify the usage and scaling of code, there are open source tools such as **Kustomize**, **ArgoCD**, and **Helm**, to name a few, that introduce a way to manage application configuration using templates. Though parameterization solutions are easy to implement, they are usually effective only on a small scale. Over time, parameterization templates become complex and difficult to maintain. Using a combination of these toolsets helps to effectively implement GitOps and continuous delivery for Kubernetes infrastructure and applications.

Kustomize extends the purely declarative approach of Kubernetes configuration and is natively built into the Kubernetes command-line interface, kubectl. Kustomize provides the ability to extend the base configuration and add, remove, and modify any delta that is usually applicable to different environments of an application. This could be any of **Role-Based Access Control (RBAC)**, network policies, life cycle hooks, scheduling constraints, tolerations, and so on.

Policy as Code

Throughout *Chapter 6*, we emphasized the importance of automation in helping us control and manage the security posture of our hybrid cloud architecture. Policy as Code refers to the use of code to manage rules and conditions around the management of security. After the determination is made for appropriately securing the hybrid cloud architecture, the policies are then written using a programming language such as Python or Rego, and this depends on the tool that is used for the enforcement of the policies. As soon as we have an asset as a representation of code, we are then able to use software programming best practices such as version control and modular design and easily extend this to the governance of cloud resources. Enterprises in regulated industries deem compliance violations to be very expensive and time-consuming to deal with. Hence, these approaches are proactive and will save them millions of dollars spent in remediating these violations and adversely impacting their brand. Some of the advantages of using Policy as Code are as follows:

- Integrate security best practices and checks into the DevOps tools and processes

- No longer required to hardcode the security policy within software code

- Remove manual steps and extend automation to the implementation of security

The following figure shows how the policy engine works as part of the policy controller. This process helps to centralize the enforcement of a security policy throughout the hybrid cloud architecture:

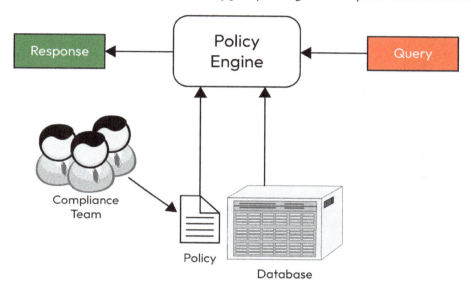

Figure 7.8 – Workflow of policy controller

One example of Policy as Code for implementing security policies in a Kubernetes environment is the **Open Policy Agent** (**OPA**), an open source engine that uses Rego as the programming language. It also uses a declarative construct that can be used to perform validations in the context of organizations' security and compliance requirements.

There are higher-level and advanced security solutions providing graphical governance of policies for Kubernetes, such as Red Hat **Advanced Cluster Security** (**ACS**) and Palo Alto's Prisma Cloud. Both of them (and probably others) offer the possibility to integrate OPA logic and be integrated into automation chains while providing useful dashboards, alerting, and risk response mechanisms.

DevOps

DevOps is the set of processes and practices that is in place to help organizations to enable software development and IT operation teams to collaborate and work seamlessly with well-defined roles and responsibilities. DevOps practice is both technical and cultural. The DevOps maturity of an organization is defined by its ability to deliver changes to applications and services at high velocity and continue to evolve by incorporating the feedback received from those applications and services running in the production environment. Here is a diagram that answers the rhetorical question of how to implement DevOps for a hybrid multi-cloud by comparing DevOps to an elephant:

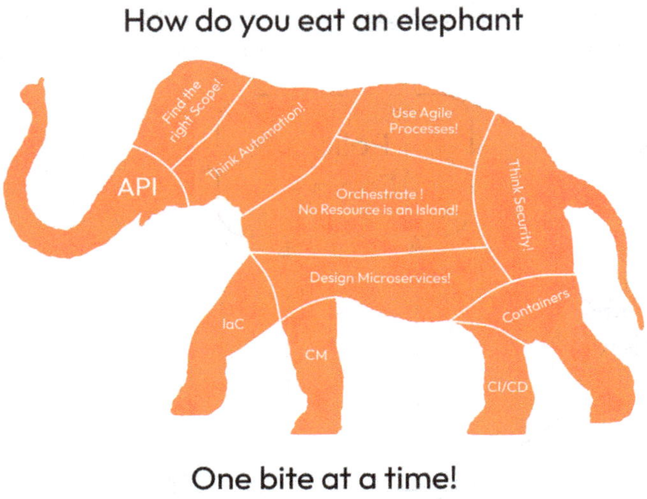

Figure 7.9 – DevOps for hybrid multi-cloud – One bite at a time

It is no secret that the advent of containers requiring immutability has clearly thinned that boundary between developers and operations. Kubernetes (being the leading platform for container orchestration) has clearly emerged as an essential tool and platform for the easy implementation of DevOps. In the previous chapter, and particularly in this section, we have emphasized the need for automation to manage a scalable and complex hybrid cloud architecture.

DevOps is essentially the overarching automation and governance entity in our architecture that helps us orchestrate the entire automation framework throughout our hybrid cloud architecture. Organizations have the utmost flexibility in choosing their tools at various phases of DevOps implementation. The many facets of DevOps include the following:

- Software configuration management

- Change management

- Requirement management

- A software delivery pipeline

- Environment management across the delivery pipeline

- Continuous integration

- Continuous testing

- Continuous deployment

- Security checks

- Continuous monitoring

- Operations and **site reliability engineering (SRE)**

Automation is the key to high-quality and highly available applications, and DevOps is the framework that helps to bring all the various facets of artifacts together in a structural manner. All the best practices discussed so far for the creation, management, and security of the hybrid cloud can be implemented, maintained, and governed using the principles of DevOps and various industry standard tools and solutions available on the market. The following figure illustrates the benefits of DevOps automation:

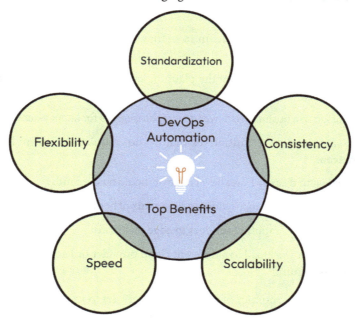

Figure 7.10 – Top benefits of DevOps automation

We focused on how to develop and deploy a hybrid cloud architecture with an emphasis on all best practices. We're next going to understand the economics of the cloud and combine all the concepts and best practices that we have learned thus far to pick a cost-effective architecture that meets all our goals and objectives. We are also going to explore the fine balance between cost efficiency and security.

What do we mean by the economics of hybrid cloud?

So, what is **cloud economics**? It is the study of cloud computing that estimates the cost of running an application on the cloud. This is particularly tricky when it comes to hybrid clouds. Organizations can have parts of the application or the entire application running on public clouds, depending on the use case. We have to take into consideration the **return on investment (ROI)** as well as the **total cost of ownership (TCO)**.

Business case

The first step in tackling the challenge of the economics of hybrid cloud is to clearly understand the business case that is driving the need. This is followed by the planning phase, evaluating the impact, and finally, budgeting to execute the plan against the business case. We will explore a few business drivers for the hybrid cloud:

- Evaluate the resource requirements of the application that address the business need. Next-generation workloads, which are focused on intense analytics and artificial intelligence, require limitless compute.

- The overhead inherited by the IT team to manage the entire hardware life cycle of the next-generation infrastructure.

- Changes in business needs outside of the planning cycle that require spending that exceeds the budgets allocated.

- The quick expansion of a business in newer geographic regions for latency-sensitive applications.

- Fail-fast, fail-safe, innovative initiatives that do not require the entire life cycle of the infrastructure team.

- The need to create and destroy lower-level environments on demand.

- The need to modernize parts of applications to have a mix of both legacy and modern components.

- The need to expand the computing services to edge platforms.

- The exit cost to move a workload to a different cloud provider or back to the on-premises infrastructure.

Before going deeper into the economics of hybrid cloud, let's get to know the difference between CapEx and OpEx. Both of these constructs are universal and apply to any business, not just IT. When it comes to hybrid cloud, the organization already has an investment in its on-premises infrastructure and hence it's important to explore the maximum capabilities of the asset. An in-depth analysis of CapEx and OpEx would result in financial modeling, and the real data needs to be compared against industry benchmarks.

CapEx versus OpEx

CapEx is defined as the process of running a business through the acquisition of assets once and it benefits the business for several years. These could refer to the acquired real estate, air conditioners, and generators, specifically for running the IT; all the servers and equipment used to build the data center would come under CapEx. All the maintenance contracts signed that help to extend the lifetime of these assets would also come under the CapEx category.

OpEx is the cost incurred to run the business on a day-to-day basis. This could include the paper for the printer, the electricity for the building, or the domain registrations for the business website. They are consumable items that get used or paid for based on usage. These are essentials for the business but they are not an asset to the organization. In a hybrid cloud, OpEx would correspond to the public cloud where the business incurs charges based on its consumption of the resources.

The following figure shows a simple comparison between CapEx and OpEx:

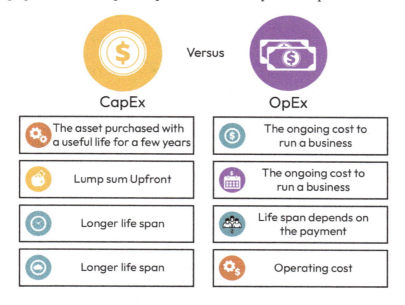

Figure 7.11 – CapEx vs OpEx

Understanding CapEx and OpEx is very important to clearly understand the economics of a hybrid cloud. This will help us determine the ROI and TCO.

ROI and TCO

ROI refers to the impact of the investment on the total business. The returns are greater than the TCO. It's very difficult to accurately estimate the ROI, particularly with a hybrid cloud because of how the tangible (efficiency and revenue) and intangible (complex DevOps processes and multiple tools) are laid out both on the on-premises side and the cloud.

The TCO refers to the total cost of investment that is associated with setting up the infrastructure. It includes the purchase, operation, and maintenance of an asset during its lifetime. Again, this gets really challenging with hybrid cloud since the cloud is a dynamic ecosystem, and estimating the TCO is a combination of cloud infrastructure, cloud migration or management, and operations cost. These have to be done individually for both the public and private clouds.

The ROI can be easily calculated using a simple formula:

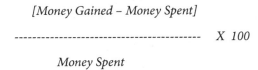

[Money Gained – Money Spent]

--- *X 100*

Money Spent

Both CapEx and OpEx play an important role in the ROI and TCO calculation because of the way they are accounted for in the balance sheet of the organization. In CapEx, the value of the asset depreciates over a period of time, while in OpEx, it is fully deductible while computing the profit and loss. Neither is a clear winner and within the hybrid cloud, the determination is carefully made after making sure the pros outweigh the cons as envisioned within the requirements of the business.

Even before we determine whether the workload is more expensive or cheaper to run on the cloud, we need to create a benchmark for comparison. What is the cost of running the workload on my existing infrastructure? This provides the basis for comparison. We then pivot to estimating the running cost on the cloud, which includes the migration, training, integration, and maintenance costs. There are also some intangible benefits we get from the cloud, such as innovation and elasticity, which need to be factored in as well.

Let's examine some of the advantages of a hybrid cloud that would tilt the economics in its favor:

- Some applications or components of the application cannot be modernized to run on the cloud. It will end up costing more to run on the cloud and will make better sense to derive value out of the existing investment in the data center. Certain databases cannot be modernized, and some legacy ERP systems are best left in the data center and become connected as part of a hybrid cloud ecosystem.

- With Kubernetes, clusters can be directly created out of bare metal servers, negating the need for hypervisors for virtualization. Using its ability to run applications as containers and dynamic scaling capabilities, the compute resources are more effectively utilized. Furthermore, since container applications are easily portable, cloud burst scenarios can be conceived for additional on-demand resource requirements, transferring the burden to OpEx.

- A hybrid cloud can also be used in a transitionary state, to prove that a cloud model can be beneficial while a complete cloud adoption is analyzed. To explain this in detail, a cloud transformation is a journey. Not all components of the application can be modernized at the same time. It would make greater sense for the frontend components of the application to scale using the cloud resources while the backend components reside in the data center while efforts to modernize them are in progress.

- There is a need for the business to react to competition immediately, putting a squeeze on the planned budget and the time it takes to go through the entire IT service management life cycle.

- When engineers use their business credit cards to spin up application resources in the cloud for testing against the excuse of agility, that phenomenon is called **shadow IT**. Hence, the use of public clouds embracing a hybrid approach is more economical than curtailing the use of public clouds.

- Compliance and GDPR requirements expect the business to continue maintaining infrastructure for its data in certain regions.

- You can do more with the same number of IT resources by transferring to an IaaS model with the addition of the cloud to the on-premises infrastructure.

Ultimately, it's the business case that determines the pattern of cloud implementation. Oftentimes, as we have seen throughout this chapter, there are different patterns of hybrid cloud available, and the choice eventually comes down to the key requirements of the workload. Cost and the business have to be delicately balanced as all choices have their advantages and disadvantages.

Various pitfalls of hybrid cloud

In this section, we are going to look at the disadvantages of a hybrid cloud. Throughout the previous sections, we looked at all the business drivers that made sense to operate in a hybrid model. However, it's imperative to look at the challenges or pitfalls of the hybrid cloud as well. As enterprises plan their implementation of the hybrid cloud architecture, they should factor in the elements discussed in the following subsections.

Complexity

In the *Architectural considerations* and *Development and deployment of hybrid cloud* sections, we looked at different hybrid cloud design patterns and also analyzed some of the business cases where a hybrid cloud would make a lot of sense. However, in all those use cases, the nature of the architecture introduces newer challenges when we need to extend the data center to include one or more public clouds.

Here are some of the challenges that practitioners contend with as they embark on a hybrid cloud journey:

- Inherently, the hybrid cloud introduces many external components that need to be part of the management and governance of the enterprise.

- Cloud economics is complex and traditional IT teams might be blindsided by the huge cost if it is not managed and governed properly.

- The management complexity spans the network, security, and several IT processes such as backups, disaster recovery, distributed storage, and management to list a few.

- IT service management principles and practices must evolve to include new cloud components and edge components if applicable.

- Processes need to be consistent when a multi-cloud strategy is adopted as part of a hybrid cloud.

- Creating a consistent interface for observability also becomes very challenging for adhering to compliance and governance.

- You cannot natively run applications on the cloud without applying modern cloud-native constructs and expect cost efficiency. Hence, applications, or the respective components, must be refactored to run on the cloud.

- Extending capabilities to accommodate business continuity guidelines further adds to the complexity.

- All these require upskilling and training the resource pool, developers, DevOps engineers, and SRE members.

- It's expensive to reverse the decision to include the cloud in architecture. Hence, decisions need to be well thought out before execution.

The following figure shows some of the tasks associated with cloud management across various categories:

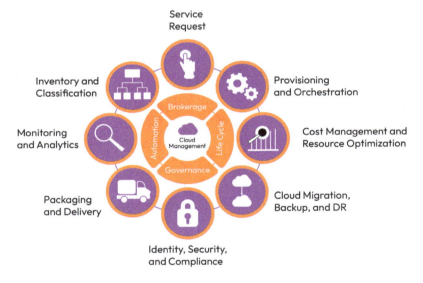

Figure 7.12 – Tasks associated with cloud management

Implementation

A hybrid cloud usually starts with the inclusion of a public cloud provider and then may slowly expand to a multi-cloud strategy. This is usually to take advantage of the best capabilities offered by each of the cloud providers. Enterprises also focus on minimizing their lock-in to any one particular public cloud vendor and use competition to get better pricing and discounts. Immaterial of whether it's a single public cloud or multi-cloud, a hybrid strategy requires additional expertise when it comes to implementation:

- Without the correct architecture, you cannot effectively leverage the cloud capabilities

- Compute, network, and storage have been extended beyond the data center

- New service endpoints that are more dispersed and may require integration with legacy applications

- Cloud migration

- Newer environments

- It brings forth the most recent and up-to-date technology stack, which may not adhere to traditional processes and usually disrupts workflows

- Cumulate logging and monitoring

- Different approaches to WAFs with the network perimeter further expanded

- Expand the automation framework to include the public cloud and newer Infrastructure as Code scripts

- A focus on all the security best practices discussed in the previous chapter

Here is a diagram that summarizes the steps around the implementation of a hybrid cloud:

Figure 7.13 – Steps around hybrid cloud implementation

Network connectivity

Network connectivity is the glue that brings all the infrastructure footprints together and makes the entire architecture work as a single application. In previous chapters, we understood how to orchestrate a container-based application using Kubernetes. Exposing the application running inside a cluster to external users is part of the networking, but the application's requirement to seamlessly talk to all other components in the architecture deals with the overall network connectivity. Hence, this layer becomes a big challenge in a hybrid environment. The following are the key elements while we design network connectivity:

- Providing access to the application to end users
- If there are multiple services running in the Kubernetes cluster, then providing the ability for the services to talk to each other
- Providing the ability for application components to talk to any third-party applications or backend databases that might exist elsewhere
- Providing the ability for edge devices and endpoints to push data if any data ingestion pipeline is part of the application

The following figure shows a simple illustration of different networks coming together and shows the separation of public and private network boundaries in a hybrid cloud architecture:

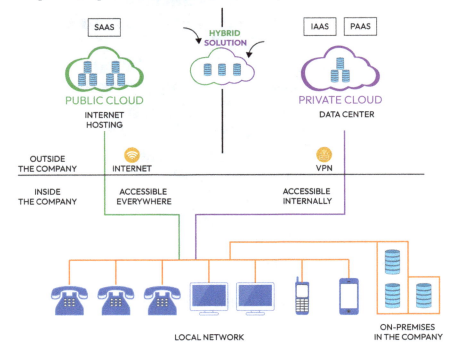

Figure 7.14 – Hybrid cloud network illustration

Across all these elements of connectivity, the key is to maintain network security. Within an on-premises data center, a public cloud, or an edge location, the basic networking for the infrastructure, platform, and application components is private. The challenge of connecting to these private networks is to do so securely. Typically, this could mean the following, depending on the architecture of the application and which footprint we use to expose the application to the end user:

- Securely connecting edge networks to a public cloud or data center. A very common use case is a data ingestion pipeline either running on the cloud or on-premises.

- Connecting applications to backend data systems running on a different cloud.

- Connecting applications to backend databases running inside data centers. Typically, these are databases that are legacy and cannot be modernized.

Let's look at the different options that are now available to make these private networks talk to each other. Pivoting back to all the network security aspects we discussed in the previous chapter, the connection can happen in a few different ways:

- If this connectivity is going to happen over regular internet, then this needs to be over a VPN connection. In this type of connection, there is a gateway component on both sides that provides a connection pipe where all traffic is encrypted in transit. Since VPN connectivity is happening over the internet, it supports all three types of hybrid connectivity requirements mentioned previously. However, there is also a key limitation that needs to be highlighted. The maximum bandwidth for VPN connectivity ranges from 3-5 Gbps. The network latency and application tolerance should be taken into consideration when choosing this option.

- However, if an extra layer of security is required coupled with higher bandwidth, then cloud providers offer ways to make a physical connection between the on-premises network and the cloud network. This provides the ability to transfer large amounts of data between the two networks and can sometimes work out to be more cost-effective than purchasing additional bandwidth with an internet service. The physical network connection is usually made through a co-location facility where the two networks come together. Even though the bandwidth is between 100 and 200 Gbps, these connections are extremely expensive. Cost is usually the prohibitive factor unless the business demands this requirement.

- There is a third option and this is a trade-off between the high cost of a dedicated interconnect and the low bandwidth of a VPN connection. This is an interconnect option that is offered by the cloud provider's partners. The connectivity between the on-premises network and the cloud provider is through a service provider who has direct connectivity to the cloud provider's network.

Between the three options that are available, hybrid network connectivity can be achieved but can be a real bottleneck in the architecture, especially when the cost and bandwidth requirements of the application cannot be balanced.

Security certification

Having analyzed the security best practices in detail in the previous chapter, we are just trying to understand the complex nature of implementing security across multiple platforms (read this as the cloud, edge, data center, etc.). If the business belongs to a regulated industry, there are considerable security certifications that may be required. From an enterprise standpoint, a proper governance process should be in place:

- Constantly review security policies
- Make appropriate risk assessments
- Identity potential vulnerabilities

The data in the hybrid cloud architecture flows through many systems. The compliance and regulatory requirements need to be maintained throughout the life cycle of the data. It needs to be protected in transit, at rest, in memory, and so on. Data gets stored in many segmented storage systems, and hence the utmost care must be taken to protect these data storage locations. The common likelihood is the man-in-the-middle kind of attack, as it flows through different cloud and on-premises environments. Encryption should be tactfully used to conceal data from unauthorized users.

The following shows the different aspects of workload security across three pillars:

Figure 7.15 – Workload security components

As the enterprise strives to maintain certifications, standardization of processes becomes a requirement to enforce governance at scale. Hence, management through automation and code becomes the only way to attain the security posture recommended through best practices.

Observability

Observability for the hybrid cloud refers to the mechanics of getting end-to-end visibility of operations related to the entire stack of the application spread across multiple infrastructure footprints. Modern applications that are based on Kubernetes are distributed by design, and hence deep visibility is very important to manage and maintain them proactively.

With the advent of microservice-based architecture and the distributed nature of application components with Kubernetes, it is very important to get deep visibility of service-to-service communication. Workload observability can be derived from different pillars:

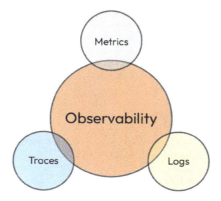

Figure 7.16 – Pillars of workload observability

The new brand of engineering that deals with managing the availability of the service is called SRE. Observability is an integral part of that. The reason why this is mentioned under pitfalls is because of how complex it gets to set up a monitoring system for the hybrid environment. Let's look at a few reasons why this is difficult:

- The enterprise has already committed to and invested in the monitoring tool stack
- Cloud platforms have good integrations with their native monitoring stack
- The are individual tools for metrics and logging
- Edge monitoring could require a different set of tools
- Building integrations with different monitoring tools to aggregate logging and monitoring data
- Diverse systems requiring different monitoring parameters that are managed by specific tools
- Massive amounts of data generated by many systems

Designing a single observability solution should go beyond conventional monitoring that incorporates all the SRE principles. It needs to integrate data from different IT systems, spanning across many layers that include networks, storage, servers, clusters, databases, and registries. The goal is to optimize the performance of the entire hybrid cloud architecture, ensure high availability of services, and provide a good alert mechanism and options to further integrate with existing automated workflows. There is a lot of data generated by a lot of these hybrid cloud systems that, if analyzed properly, can increase the predictability of the health and performance of the system and thereby improve productivity immensely. Finally, to get the best results out of the monitoring systems, it's key to completely automate, and that becomes a considerable challenge in extending the automation framework to monitoring.

Cost

We have finally come to the most important consequence of all the challenges that we have discussed under pitfalls: cost, which inherently becomes high because of all the implementations around the complex hybrid cloud architecture and the best practices associated with it.

Here is a list of all the hybrid cloud cost considerations:

- Management and support fees
- Data transfer
- Storage fees
- Hybrid cloud software and server fees
- Customization and integration fees
- Compliance fees

Finally, it boils down to whether or not the business can afford the cost of a hybrid cloud. Hybrid cloud architecture is chosen when the benefits outweigh the cost. There is a combination of both OpEx and CapEx, and we discussed the economics of a hybrid cloud in detail, which means there is enough efficiency to be realized as demanded by the workload and business.

Summary

In this chapter, we understood the architectural considerations for a hybrid cloud, followed by the economics of a hybrid cloud to understand the impact of cost-managing both a data center and one or more cloud providers. Here are some important considerations as you develop a hybrid cloud strategy:

- Understand the business objectives clearly and develop a strong cloud strategy
- Implement a cloud strategy, whether single or multi-cloud, that provides flexibility in terms of portability and consistency
- Choose solutions that are based on open source technologies
- Revisit the compliance requirements and map them to the hybrid cloud strategy

- Strive to have unified IT management across all infrastructure footprints and make sure all enterprise requirements are met from a security, networking, and resource management standpoint

It is very clear that the decision to go with a hybrid cloud architecture depends on many factors. Also, the final architecture needs to be well thought out and analyzed against the business requirements of the enterprise. There is no easy answer but hopefully, the knowledge gained from reading this book helps you to make an informed choice.

Hope you enjoyed reading the book!

Index

Packtpub.com

Subscribe to our online digital library for full access to over 7,000 books and videos, as well as industry leading tools to help you plan your personal development and advance your career. For more information, please visit our website.

Why subscribe?

- Spend less time learning and more time coding with practical eBooks and Videos from over 4,000 industry professionals

- Improve your learning with Skill Plans built especially for you

- Get a free eBook or video every month

- Fully searchable for easy access to vital information

- Copy and paste, print, and bookmark content

Did you know that Packt offers eBook versions of every book published, with PDF and ePub files available? You can upgrade to the eBook version at packtpub.com and as a print book customer, you are entitled to a discount on the eBook copy. Get in touch with us at customercare@packtpub.com for more details.

At www.packtpub.com, you can also read a collection of free technical articles, sign up for a range of free newsletters, and receive exclusive discounts and offers on Packt books and eBooks.

Other Books You May Enjoy

If you enjoyed this book, you may be interested in these other books by Packt:

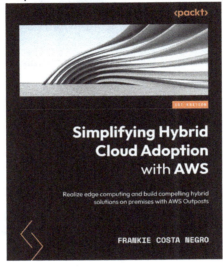

Simplifying Hybrid Cloud Adoption with AWS

Frankie Costa Negro

ISBN: 9781803231754

- Discover the role of AWS Outposts in the hybrid edge space
- Understand rack components with typical use cases for AWS Outposts
- Explore AWS services running on Outposts and its capabilities
- Select, order, and successfully deploy your Outposts
- Work with Outposts resources for hands-on operations
- Assess logical and physical security aspects and considerations
- Monitor and log configuration and usage to improve your architecture
- Maintain and troubleshoot hardware and software that run AWS services

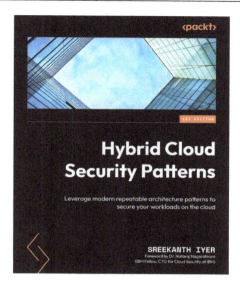

Hybrid Cloud Security Patterns

Sreekanth Iyer

ISBN: 9781803233581

- Address hybrid cloud security challenges with a pattern-based approach
- Manage identity and access for users, services, and applications
- Use patterns for secure compute, network isolation, protection, and connectivity
- Protect data at rest, in transit and in use with data security patterns
- Understand how to shift left security for applications with DevSecOps
- Manage security posture centrally with CSPM
- Automate incident response with SOAR
- Use hybrid cloud security patterns to build a zero trust security model

Packt is searching for authors like you

If you're interested in becoming an author for Packt, please visit `authors.packtpub.com` and apply today. We have worked with thousands of developers and tech professionals, just like you, to help them share their insight with the global tech community. You can make a general application, apply for a specific hot topic that we are recruiting an author for, or submit your own idea.

Share your thoughts

Now you've finished *Achieving Digital Transformation Using Hybrid Cloud*, we'd love to hear your thoughts! Scan the QR code below to go straight to the Amazon review page for this book and share your feedback or leave a review on the site that you purchased it from.

`https://packt.link/r/183763369X`

Your review is important to us and the tech community and will help us make sure we're delivering excellent quality content.

Download a free PDF copy of this book

Thanks for purchasing this book!

Do you like to read on the go but are unable to carry your print books everywhere? Is your eBook purchase not compatible with the device of your choice?

Don't worry, now with every Packt book you get a DRM-free PDF version of that book at no cost.

Read anywhere, any place, on any device. Search, copy, and paste code from your favorite technical books directly into your application.

The perks don't stop there, you can get exclusive access to discounts, newsletters, and great free content in your inbox daily

Follow these simple steps to get the benefits:

1. Scan the QR code or visit the link below

https://packt.link/free-ebook/9781837633692

1. Submit your proof of purchase
2. That's it! We'll send your free PDF and other benefits to your email directly

www.ingramcontent.com/pod-product-compliance
Lightning Source LLC
Chambersburg PA
CBHW080522060326
40690CB00022B/5002